PRAISE FOR
COLLAPSE

"My experience is of a book with a deeply compassionate intention, to reach those that need this guide right now . . . Juan Pablo offers signs and tools for our different responses; this being a journey, with complexity and nuance, with inner and outer implications."
—GAIL BRADBROOK, co-founder of Extinction Rebellion

"If you're new to the global discussion about civilizational collapse, this book will acquaint you with the best thinking on the subject. Collapse is scary to think about, but if you're intelligent and paying attention to the world around you, it's an unavoidable subject. Juan Pablo Quiñonez is an informed, kind, and thoughtful guide to why civilization is coming apart and what to do."
—RICHARD HEINBERG, Senior Fellow, Post Carbon Institute, author of *Power: Limits and Prospects for Human Survival*

"Collapse is a prescient book that confronts the feel-good delusions of endless growth and technological salvation . . . this book creates space for personal reflection leading towards a deeper understanding of the roots of our predicament. Read it, feel the grief, and find the courage to act."
—DERRICK JENSEN, author of *A Language Older Than Words* and *Endgame*

"Contemplative analysis of civilization's potential for collapse, with holistic answers."
—BOOKLIFE BY PUBLISHERS WEEKLY

"Quiñonez's analysis of global crises . . . will appeal most to those who already agree with the author's assertions."
— KIRKUS REVIEWS

COLLAPSE

NAVIGATING CIVILIZATION'S PREDICAMENTS WITH WISDOM AND COURAGE

JUAN PABLO QUIÑONEZ

BOREAL
CREEK
PRESS

ISBN 978-1-7772838-2-7 (trade paperback) | ISBN 978-1-7772838-3-4 (ebook) | ISBN 978-1-7772838-4-1 (audiobook)

Copy edited by Brock Peters
Cover design by Christina Wulandari
Figures and illustrations by Juan Pablo Quiñonez

This book is available at quantity discounts for bulk purchases. For information, please email sales@borealcreek.press.

Contents

Part 1: Prep Work

1. Into Unknowing

This book offers readers an understanding of the converging social, economic, and environmental crises we all face, while recognizing that absolute truth is beyond reach. It's for those who wish to perceive the interconnections of our challenges and explore perspectives that may help us respond in our own unique ways. The subject of *collapse* is powerful. Depending on how it's taken, it can be helpful or toxic.

In this book, I invite you to explore and engage with a difficult subject without falling into the pitfalls of toxic positivity or total pessimism. If you are comfortable existing in a state of not knowing, of uncertainty, while still moving forward according to your own compass, this book is for you. This is an opportunity to grieve, acknowledge our losses, and contemplate how to respond. It is an analysis of our global situation that does not seek solutions for maintaining our current way of life but focuses on facing and responding to this century's potential for profound loss, transformation, and opportunity.

Collapse is divided into three parts: Prep Work (Chapters 1–3), Predicament (Chapters 4–9), and Responses (Chapters 10–12). The first section brings awareness to common mental, emotional, and cultural barriers towards the subject. The middle section explores what *collapse* is, why there is no "fix," and the conditions that led to this predicament. The final section suggests some responses and helpful ways to approach the future.

Many people may not be interested in collapse, but collapse is interested in humanity. We live in apocalyptic times. Recalling the ancestral meaning of that phrase, these *revelatory* times uncover what has been hiding in plain sight. For thousands of years, various entities have been at play—narratives, groups, and forces—herding humanity towards this moment of reckoning. This pushing and pulling has mostly taken place outside our conscious awareness. The changes that are emerging now are dynamic, and any attempts to control them will fail. The only way out of this wild ride is through.

Our individual and collective shadows have been ignored and suppressed for far too long, and now they are bursting out. Around us, most of humanity has traveled down a certain singular path. This path is self-reinforcing: the further we go, the more resistance there is to taking a different direction. Naturally, along this path exists a point of no return—and from multiple angles, it looks like we have crossed that point.

What is the true face of humanity? Can we look at ourselves without idealizing or condemning who we are? Can we observe with equanimity what repulses us or what we aren't aware of?

There are those who see, those who see when shown, and those who don't want to see. This book is not for the latter. Not everyone is ready to ingest this powerful medicine. This journey is neither quick nor easy, and it may lead to a spiral of understanding and uncertainty, maturity and grief, agency and accountability. But it's a much-needed rite of passage in these times of unprecedented change. If adequately digested, there's a certain liberation and understanding to be found, but also a shared responsibility to bear.

As a collective, humanity will undergo this rite relatively soon. But those of us who can embark on this journey ahead of time may be empowered to help guide others as they become ready, allowing them to find their own paths without making the usual mistakes of propagating counterproductive mindsets, ideas, and actions.

Reading this book will hopefully give you a more realistic perspective on what's happening. Making wiser decisions is a benefit of having a view that better aligns with reality. But the best reason for undertaking this journey is to honor the times we are living in. At this moment, everything is on the cusp. What we do or don't do matters a lot. And the stakes are as high as they can get. It is urgent, yet it can't be rushed. Jumping hastily into action will be counterproductive.

In this book, I offer a perspective on why we are experiencing such an extreme imbalance between nature and humanity, as well as on the complexity of the metacrisis (i.e., the underlying crisis driving the interconnected global crises, also called the *polycrisis*). The final chapters share broad suggestions on how we could respond in wiser ways. This book is for anyone who has felt or suspects that we are going in the wrong direction at an increasing pace, and for whom the reasons to believe otherwise are fading away.

If you believe, without the slightest doubt, that modernity and civilization are invulnerable or immortal and that their current challenges are just minor obstacles, this book is not for you. However, if you suspect that humankind is doing its best to tackle the converging crises with its current tools, perspectives, and challenges, and that our default responses are not working, then perhaps you are ready to consider that things may not go as humanity expects. Life happens, after all.

To help you decide if you should read this book, consider how many of the following statements strongly resonate with you:

- You put considerable effort into making people see you as you want to be seen.
- Uncertainty makes you feel very anxious.
- Considering multiple options makes you feel overwhelmed and immobilized.
- The lack of concrete solutions to the world's problems makes you feel hopeless.
- Science and modern technology hold the answers to our global problems.
- We already have solutions to our global issues and just need the will to implement them.
- There are groups of people who have all the answers.

If you agree strongly with the sentiment in most of those statements, this book may not be for you at this time.[1]

If you are heavily invested in the existing system of business as usual, it might be challenging to see how things really are; you may not be prepared to examine the contradictions in your assumptions.

If you read a substantial amount of this book and feel you can't continue, I suggest you complete it anyway, at your own pace, and maintain some distance from what doesn't resonate with you. Partially reading it might get you started on this journey, but you may find yourself in a state of fear or despair without clear guidance on navigating those challenging stages in a way that isn't toxic. There's a list of peer-support resources at the end of this book. Proceed with caution. If this book isn't right for you at the present moment, please give it to someone else.

I invite you to explore the roots of the polycrisis, its magnitude and complexity, and the changes that may need to take place in us to respond wisely. This book suggests responses to consider and responses to avoid. It's essential not to react in a way that causes more suffering.

In writing this, I acknowledge that what I perceive is filtered by my nature and nurture—who I am and how I was raised—and that some biases that I'm not aware of also influence me. I don't know for sure what is happening or what will happen. This "unknowing" puts me slightly at ease and allows more clarity to pass through. The mind of an expert believes it knows everything and thus is rigid, but a "learner's mind" is open to what is.

Throughout this book, I share a thoroughly researched perspective on where we stand collectively, as well as relevant ideas I've encountered throughout my life. Some of these ideas have lingered in my mind and—mostly through unconscious processes—transformed into insights. Some of those insights were further clarified and strengthened by my own observations.

How could anyone write about this complex topic in a way that doesn't lead to responses which cause added suffering? No one can know when they set out how their writing will be received; this is a humbling lesson for all writers. I had no option but to let go and trust the unfolding. I strove for self-honesty, to maintain a healthy balance of attachment and detachment, to be present in the process rather than seeking specific outcomes, and to be grounded in the miracle and sacredness of life.

The origins of the universe and of life are surrounded by mystery, but there's no shortage of stories. The biological evidence points towards a long, winding path of miracles, adversity, grit, evolution, diversity, and change.

Life's emergence was facilitated by the inherent principles of the universe. There are many names for these forces and laws, but often these names carry the baggage of old dogmas. Life is emergent and self-organizing, and it has a fractal nature. It doesn't possess a central command. Life taps into the inner ancestral wisdom that connects it to its source. There are no systems of governance; the life force itself is self-governing. Life is fractal in that it's cyclical, as in a spiral. It is always in flux because reality itself is in flux. There's also a repetitiveness to life—to an extent, its parts resemble other parts: as above, so below; as within, so without. Life has a nesting-doll aspect.

The same forces that make physical objects go up and down, acting through energy and matter, also act on life. Life plays by universal rules. The dance emerges from harmony. We may notice patterns in the dance, through scientific methods or plain observation, but there will always be a high degree of mystery. However, there is a purpose, a directional force traveling alongside the arrow of time. This "purpose" is trending towards increased complexity in the expressions of life and increased inter-relationships.

We are the dust of the stars and the rays of sunlight shining on our skin. We are the soil where plants grow. We are the oxygen inhaled and the carbon dioxide exhaled. We are the water flowing through our own veins and within each of our cells. We are the energy tingling in our chests and limbs. We are the present moment. We are our ancestors and our descendants. We are the ripples left on this Earth, and thus, we are the Earth.

The insights of non-separation or interbeing help us to see the Earth as a self-organizing superorganism—Mother Earth or Gaia. People from all walks of life have perceived this insight. For any given planet, what are the chances of having clean, drinkable water falling from the sky, moderate temperatures and seasonal cycles, and a sun to energize the plants and animals? The Earth, with its magnetic field, protects itself from cosmic radiation as it orbits the sun at an unimaginable speed, and its atmosphere maintains a fairly stable temperature. How did beautiful sandy beaches or awe-inspiring mountains come about? The gifts of air, land, water, and the sun shouldn't be taken for granted. Our ancestors warned us against that.

I recently stumbled upon a dark impulse present in our shared reality. I felt and saw its signs a long time ago, but it's only now that I've connected the dots. It is not the impulse for death, since life and death are two sides of the same coin. No, this impulse is something else. It says "No" to life. This impulse is a negation of life, of being an embodied organism and all that comes with it. This force seeks the void, annihilation, and extinction. It's the voice saying "Jump!" when we are at the edge of a cliff. And this impulse has taken hold of a portion of humanity, leading it to undermine its own foundations. Yet this impulse is a symptom, not the cause.

This impulse is here for a reason. It's a natural force. We are caught in the interplay of creation, destruction, and extinction. As localizations of the cosmic whole, we are choosing our fate at every moment. Do we *consciously* choose to be creators of reality, helping to shape the unfolding of life despite the challenges we face? Or do we let these mostly unconscious impulses keep us on a path towards the abyss?

How do we face reality without looking away? How do we ground ourselves in the sacredness of life? How do we embrace the feelings that arise in our hearts? And how do we respond in true service to ourselves, others, future generations, and life itself? These questions guided me as I wrote this book.

The Earth has gone through extreme challenges before. In the past, there were five major extinction events. And our direct ancestors, in different forms, have lived through them all—our lineage goes back to unicellular life. Some of those mass extinctions occurred in the course of unimaginable catastrophes, under circumstances that make nuclear war seem like a walk in the park. We are the continuation of those survivors' sustained efforts, pains, and joys. Despite all the adversity, they survived and shared the gift of life with us.

Among the multiple crises of our era, considering whether industrial civilization could suffer a major decline in the near future is a worthy use of our time. If it's a real possibility, it doesn't mean that all is lost and there's nothing worth doing. The moment we find ourselves in has enormous potential. As we stand on the cusp of change, this moment represents an opportunity to have a leveraged impact on our collective future.

Are you willing to go beyond your comfort zone? It's okay if you aren't. Whatever your choice, I encourage you to stay true to your own wisdom. The subject of *collapse* is best understood through gradual immersion. *Too much, too fast* can be a recipe for disaster. Take your time.

This chapter is a signpost, a moment to consider following your intuition. Pause and focus on your heart. Would you like to know what's happening and consider the possible responses? Or would you prefer not to know?

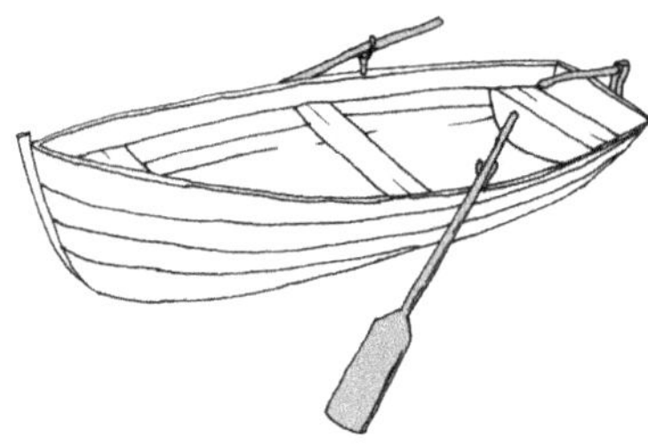

I have always perceived something odd regarding our relationships with each other and with the living world. As a child, I sensed that the way we were doing things couldn't continue forever and that there would eventually be a reckoning. That thought evolved from a deep suspicion to a well-informed insight as I connected the dots among multiple signs and trends.

For years, my understanding developed as I climbed the ladder of awareness.[2] I began by noticing one systemic issue, and I soon noticed various others. I started to understand the interconnectedness of these problems and the need for a holistic approach. Finally, I realized that these were not singular problems, but a predicament encompassing every aspect of our lives and our entire modern world.

In 2018, I spent significant time reading open-source information relevant to collapse. Since then, I've grappled further with the subject and gradually developed a more coherent understanding of it—and of the possible responses.

Metaphorically speaking, if I'm in a building and I notice a fire, I will probably warn others and try to be helpful. Some people may not believe there is a fire and will continue as they were. Others may get overwhelmed by the situation and stand stock-still in shock. Some may try to silence me because they don't see the smoke or fear that sounding the alarm will create panic. However, several people will perceive what's happening and, in turn, help themselves and others.

An urge to share what I see with others overpowered my hesitations—namely, the fear that this subject might propagate unhelpful energies and ideas. There are many rationalizations for sharing these observations, but my intuition and life force ultimately compelled me. I'm sharing alternative maps and compasses. Perhaps you might discover that some of them align with the surrounding terrain. Maybe those will provide you with some orientation.

As the author, I bring a unique perspective to this material; I've lived in a specific context with a particular nature and nurture, allowing me to be receptive to these issues while maintaining a relatively clear outlook. My mind tends towards a bottom-up analysis, and although I'm not a scientist or an academic, I have an understanding of their methods and language. I value striving to maintain an open mind rather than projecting my hopes and expectations. For years it has been my fixation to study and ruminate on this topic, and I've felt a calling to organize and share what I've learned. I'm answering that call with some hesitation and with care.

I believe this subject will soon enter the mainstream realm, which is one reason I wrote this book. The topic should be handled skillfully. Without a grounded and holistic understanding, it's easy to fall into thoughts and actions that will probably create more suffering.

I've had the good fortune of being in nature in relative solitude for long periods. That distance likely helped me see our situation from a wider view. I'm also an outdoor survival expert—perhaps some people will put me in a box and dismiss this sincere book as biased because of that expertise. I've tried to counterbalance any potential biases from my inclinations, culture, and worldview, but, like any other human, I cannot do so completely.

I did my best to ground this book in what is. My intention has been to seek what is actually happening—not to confirm my intuitions. I am not attached to an agenda or outcome other than honoring the perspective I've been given through experience, research, and insight. And I'm honoring it by sharing it with you.

My journey towards awareness of this unraveling has been gradual, shaped by personal experiences and a strong curiosity. The strong intuitions I experienced during my teenage years gradually deepened into understanding over time. However, it wasn't until later in life, when I had the energy, time, and interest to delve deeper, that my understanding matured. I voraciously sought clues in books, reports, lectures, journal articles, and documentaries—driven partly by a desire for certainty. I can't replicate such a slow, progressive approach to the topic in this book. But I can point to possible major obstacles to the awareness of this unraveling—hopefully that will help.

Throughout my life, I've learned to value humanity, the living world, respect, autonomy, freedom, truth, compassion, wisdom, love, courage, sacrifice, community, family, and the future generations of humans and living beings. I've always been immersed in multiple worlds. I was born and raised in Guadalajara, Mexico. Even though it's a big, sprawling city, I was fortunate to grow up next to a forest. As a child, I spent much time chasing lizards and snakes, and rock scrambling in dangerous terrain. I was lucky enough not to grow up in a bubble; I had contact with people across the whole socioeconomic spectrum. I am *mestizo*, which means I have Indigenous and European ancestry. My native language is Spanish, but I live in Canada. Inhabiting multiple worlds has helped me get comfortable with complexities and contradictions.

I experienced the belt-assisted upbringing common among Latinos. And I was bullied in elementary and middle school. I don't put what others believe or say on a pedestal—it doesn't matter if they are authority figures or experts. I don't have any reservations about questioning and examining ideas and assumptions to see if they hold weight. I'm interested in getting closer to the truth.

I've always had a divergent mind. I am self-taught and blunt, think a lot, and value autonomy over conformity. Systems thinking, pattern recognition, bottom-up analysis, and hyper-focus come naturally to me. I notice minor details others may miss, yet I also perceive the big picture. These tendencies are both weaknesses and strengths, depending on the context. These are the parts of my personality that I bring to this discussion.

I've also had my share of reality checks and ego erosion. At nineteen, I joined the French Foreign Legion. But my expectations were unmet, and I didn't fit in well. Throughout my life, I have faced multiple challenges that have made me examine my true desires. What was I really looking for?

After spending one hundred days in solitude in the boreal wilderness during wintertime, I started to realize that a lot can be gained from swimming against the current—moving in the opposite direction of our expectations and those of our society. In 2021, I participated in the survival TV show *Alone*, and I put that lesson into practice. I leaned into hardships with curiosity and acceptance. My body deteriorated as the weeks passed, but I wasn't resisting it. I cared for myself, doing what would keep me there for as long as possible; but, simultaneously, I wasn't fighting the reality of the situation. I acknowledged and surrendered to it without giving up. After spending seventy-eight days surviving alone in Labrador, I won.

Many months later, I had a very vivid "dream" in which I was shot. This dream felt incredibly real despite its distorted reality. I was half-asleep and felt a lot of pressure on my chest. My heart rate shot straight up. I had a powerful sensation that I was dying. My mind and body agreed that I was likely a goner. My brain briefly considered what dying would entail and reviewed the meaning of my life. I decided that it was okay if it was my time to go. In my own way, I kept my chin up. Experiences like these have helped me explore the topics featured in this book without being constrained by dominant views or needing to validate or protect my assumptions.

2. Endings Are New Beginnings

Modern civilization is entering a stage of decline. Most people are not aware of this—or they're in denial—but the evidence is emerging. In the last few decades, modernity and civilization have been undermining the foundations of our societies and the living world at an ever-accelerating pace. But nothing goes on forever. The cracks are visible if we take the time to notice. There's a lot of noise, yet there is clarity waiting to be found.

What's going on in the world? Some people are sounding alarms: There is increasing economic austerity, social media is unhinged, mental health issues and addictions are rampant, and the possibility of widespread war is not far-fetched.

Life on Earth is under increasing pressure from our population and the consumption enabled by industrial civilization. Economic issues like inflation, rising debt, and extreme inequality are rampant worldwide. Modern civilization's over-reliance on oil, a finite resource, was a concern half a century ago. Today, we are witnessing the consequences of diminishing returns on the energy used for oil extraction, with the rise of fracking, offshore drilling, and oil sands. Yet confronting the precariousness of relying on limited resources is not on most people's agenda.

At first glance, our daily lives may seem stable, but the foundations of the world as we know it are deteriorating. The better we understand what is happening, the better equipped we will be to respond. This moment calls us to look at our individual and collective shadows and see what is really there, rather than what we might wish to find. To internalize the content of this book, one must embark on the winding journey, both mental and embodied, of understanding and accepting the reality of the situation.

We've all heard similar pleas: *Our society is in trouble! There are many problems worldwide, but X is of the utmost importance. The window of opportunity to act is closing. We must mobilize to turn things around. The solutions exist, or at least we will develop them soon. If we do Y, things will get better. With enough work and political will, and by convincing certain people, we'll overcome this crisis.* These sentiments could represent somewhat familiar and comforting ways to frame a particular crisis grabbing our attention. But they are not based in a thorough understanding. What is happening is much more complex, and the vast majority of responses pushed by so-called experts are misguided. In fact, deep down, you may already know this.

We are navigating successive crises: from wars to energy shocks, pandemics to wildfires, and famines to droughts. These crises will intensify. Within a human lifetime, the most probable scenario is a gradual decline in the foundations of industrial civilization, human well-being, and the health of the living world, punctuated by major disruptive events that reshape the systems of civilization. One of these disruptions may even be identified in hindsight as the point at which a relatively sudden collapse occurred. That point wouldn't mark the "end of the world," but the end of most of humanity's current mode of existence. It would also be a new beginning. This book serves as an early warning to brace for impact. Here, we will delve into what's happening and how we can play a more conscious and active role.

The perspective that modern societies are acting from as a response to these issues reproduces and amplifies the same mistakes, forcing us into further imbalance. This imbalance will resolve itself, as it always does, but we have a choice regarding how far we will push the boundaries before balance is restored. It's up to us to choose how much suffering this will entail. There is no escape from this mess.

The predicaments described in this book shape every aspect of our lives as individuals and as part of humanity. They are the overarching themes that will continue to shape our lives and those of our families and communities, whether we know it or not. Now is the time for a more conscious participation in their unfolding.

Many of us avoid thinking about unpleasant things such as our mortality. These thoughts often threaten the stories we cling to about the world and our roles in it. Usually, worldview-disrupting ideas contradict our sense of identity and self-worth, so our egos take them as a direct attack. The ego may go into survival mode and build stronger defensive walls to block that information. This aversion to thinking about things that threaten our worldview is probably one of the biggest obstacles to building an awareness of modern civilization's endgame.

How is contemplating our death valuable, anyway? Well, it makes us wiser and helps us live more fully. It encourages us to be better-engaged co-creators of our world. It also grounds us and helps us honor the miracle of life. All those benefits can also apply to contemplation of this book's subject. If we're ready, exploring uncomfortable ideas like these can be rewarding, but we must be open-minded and willing to have unpleasant experiences at times, just like in any adventure. Getting outside our comfort zone requires courage—the type that emerges from the heart. It requires enough curiosity to overcome some of our fears and accept our challenges.

Are you curious to learn more about the unraveling of modernity, about its whens and whys? Or are you afraid exposing yourself to this idea will be like opening a can of worms—one that can't be closed again? When we open ourselves to this possibility, it can be quite an intellectual and emotional rollercoaster. It is very similar to navigating the stages of grief. This makes total sense, since in some way we're mourning significant losses. We are likely to feel fear, despair, sadness, anger, and anxiety. Yet we may eventually have greater acceptance, clarity, purpose, and freedom.

An old story tells us about a group of blind men who had never encountered an elephant. Together, they attempted to make sense of it. One of them touched the trunk and said, "It's like a snake." Another one grabbed the ear and felt some sort of fan. The one in contact with a leg described something like a tree trunk. Someone touching the side claimed the elephant was "like a wall." Another person handling the tail described it as a rope. And the person holding a tusk claimed to be grasping a firm and smooth spear. This story symbolizes how each of us can see parts of reality but never the totality.

To minimize confusion as we explore reality, we should acknowledge our limitations. We learn and are taught many things about reality. But rarely is it noted that reality won't let itself be squeezed into frames or concepts. Reality is infinite and mysterious, with no beginning or end. Reality is paradoxical. It is found in the tension of opposites, like light and darkness. It refuses to be grasped. No one can reduce reality to concepts; for example, you can't write a book that encapsulates the totality of a specific living tree. Even filling all the paper ever created with our writing wouldn't capture the life and being of that tree. Language is a set of symbols used as pointers and can never embody reality. It's a map, not the territory.

A big challenge in perceiving reality is that we are entangled with it. We rarely consider the separation that's implied when we discuss subjects and objects. Yet the distinction between a subject and an object is partly false. Everything is part of the whole. Nothing can stand on its own, separated from anything else. Thich Nhat Hanh's concept of *interbeing* hints at this. We cannot observe reality without having a relationship with it. Lacking an awareness of this relationship with reality and how the relationship shapes it, we overlook key parts of the picture.

If reality is a web in which everything is entangled, any one thing affects everything else. A full comprehension of these relationships is beyond us, yet we come to intuit many of them. We tend to see causes and effects without realizing that these stories can't capture the immensity of the entire picture. Reality is a dance, and this dance is not predictable or knowable. There is no prescribed fate. Reality is a live experiment. There's no reverse, pause, or forward—only the present moment's flux. There seems to be a spiral rhythm and an inner order, even if our limited understandings do not reveal them completely to us.

Some people attempt to see reality exclusively through the lenses of logic and reason. But would you choose who to marry based on logic and reason alone? Is life a problem to be solved, or is it about something more wholesome? If we wouldn't live our lives based on a math formula, what makes us think the universe itself would follow one? Put differently, it helps to view the universe as an infinite interconnected tangle that, in each moment, comes up with the meaning of life—but remember, we are a necessary part of the process.

Metaphorically, I am putting this message in a bottle and throwing it into the ocean. A deeper understanding of these factors, along with individual and collective transformations, would in theory lead us to respond in wiser ways. Responding wisely to this everything-crisis will hopefully minimize suffering for everyone; what we personally do about it is up to us.

Regardless of our identity, location, or profession, the situation I present here is profoundly relevant to us, our families, our communities, and our world. We are all key participants in this situation, whether we acknowledge it or not. This era of profound changes amplifies our actions and inactions, echoing them into eternity.

This century we face a series of challenges arranged one after another. Only some of us have noticed our situation and connected the dots. We were born in the times leading up to an enormous rebalancing of humanity. We didn't choose to be here, but these are the cards we were dealt. The story doesn't end here. The stage has been set. And we are now playing our respective roles. I invite you to play yours passionately, whatever it may be.

3. Clearing the Fog

Before exploring the polycrisis in detail, a veil must be lifted. Without laying the groundwork, this medicine can be hard to swallow and may not have a healthy effect. And even despite this prep work, it may prove difficult to fully digest the insights we're exploring. Stepping out of our comfort zone with openness and courage is required. This chapter explores some mental and emotional barriers that stand in the way of perceiving what's happening and provides context to help us see through those barriers. The following sections suggest ways to approach the ideas in this book; reflecting on the thoughts and emotions they stir in us will be helpful.

I've noticed that many people who have realized the magnitude of the situation have experienced some sort of ego death, reality check, or similar phenomenon. They have died symbolically but are physically still alive. Deeply considering our mortality or possible disablement can have a similar effect, and maturity and clarity may, in turn, arise. Many ancestral rites of passage roleplay this journey, but they are rare nowadays. These experiences can be described as the shedding of boundaries and attachments, and opening up to what *is*. An ego erosion isn't strictly necessary for perceiving our situation, but it does weaken many of the barriers to perception.

Do you remember changing your mind about things you'd believed firmly in over the past five years? If you can remember doing so, perhaps consider holding on to your present beliefs more lightly. I don't suggest that you abandon your beliefs or change your mind, but only that you create some distance—an opening.

Tending the Boat Within Us

"Tending the boat within us" is a visualization tool—one I discovered in my reading and have modified—that can help us connect with our own complexity and emotions.[3] It will be a helpful tool for us, because this book's subject has many layers. If we cannot sit with complexity within ourselves, we will find it difficult to sit with complexity on the outside. Another good reason to use this tool is to attend to our emotions while reading. Sometimes the topic of *collapse* can cause strong reactions in us. This energy can build up and burst out if we don't acknowledge it and let it be. If we don't pay attention to these thoughts and emotions as they arise, we can become defensive and fail to engage fully with the topic; instead, we risk withdrawing from it.

Imagine there's a rowboat within you; its passengers are the voices, energies, and thoughts that appear in your body and mind. For instance, among those passengers could be a parent's or a spouse's voice, or a recurring emotion or thought. Some passengers are loud and frequently jockeying for control of the tiller; others are distant and only come close to the tiller when they seek to wreak havoc. Multiple passengers may hold the tiller simultaneously, and some passengers row backwards or even rock the boat. The same few passengers often hold the tiller, but occasionally others take over.

As you continue reading this book, I will suggest that you tend your boat by prompting you with the symbol below. Consider it a signal to take a moment to acknowledge what's happening with the passengers in your boat and to let them be. If the passengers are seasick or the signs of a mutiny appear, it may be time to pause and reflect.

Initiation Into Adulthood

Most modern societies don't have rites of passage into adulthood. Thus, the adolescent paradigm persists in many adult-age people. The transition from adolescence to adulthood is an initiation into our life's mission and gifts. It also invites us to become better aligned with creation. This process involves vulnerability and restructuring ourselves, so guidance from elders is essential. Unfortunately, most modern societies don't facilitate these initiations.

Properly initiated adults have a greater capacity to engage with the world fully. They are better able to participate in an interdependent relationship with their community. An initiation, like an illness, an accident, or an intimacy with death, also helps us face our own shadows. These shadows are aspects of ourselves that we reject, repress, or deny, and an initiation can bring these to our awareness.

We should all learn to live with our shadows without making them our enemies. Repressing them has a counter-effect that may cause them to burst out unexpectedly and can lead to suffering for ourselves and others. An improved self-awareness and self-understanding will not solve our problems, but it helps us perceive their roots and dynamics better.

The inner journey is about letting go of old paradigms; it takes great courage to do so. Attempting to have our affairs in order is a courageous and powerful act that can improve the world. Regardless of how we got here collectively, this era requires us to grow up and stand up, because many adult-age adolescents are burning the village to feel its warmth.

Shadows of the Eternal Child

"No tree can grow to Heaven unless its roots reach down to Hell." Carl Jung, *Aion*

Many of us have experienced phases of life when we thought we had all the answers. This often occurs in our younger years. However, as we grow older, life demonstrates to us that we don't have everything figured out. But many of us are, in some sense, still stuck in that earlier place, with our minds full of ideas and stories. We lack the cultures, mentors, and rituals to help empty our minds and fill them with something more fresh and generative.

The media keeps pumping out films about superheroes as if there's no tomorrow. These superhuman heroes always manage to save the world. They don't have a problem with the collateral damage from fighting their monsters—you gotta break some eggs to make an omelet. They know they are the good guys, and they recognize the baddies at first glance. Though the bad guys are often defeated, there's always a comeback, a desire for revenge, and the battle starts again. It's a never-ending cycle inside a black-and-white world.

What if the path to a true breakthrough was within us? What if we sacrificed the righteous hero in ourselves rather than sacrificing the villains outside? What if we turned our attention to our own limitations, our kryptonite? "This is madness!" our internal hero yells. This shift challenges the very core of our identity.

It takes immense courage to dethrone our righteous hero, our personal grandiosity. It's a difficult sacrifice because it may initially lead to doubt and confusion. Imagine living your whole life with black-and-white vision and then suddenly gaining the ability to see in color. At first, this would be disconcerting. Your brain would undoubtedly take some time to adjust to the increased complexity of the signals—but you'd be the better for it.

A greater understanding also opens our hearts to shocks and suffering. Where's the benefit of that? Learning to live with our limitations—sacrificing the righteous hero—allows us to rise above our weaknesses. Once we begin to see our deficiencies, the path to redemption appears. Rebalancing ourselves is how we heal. And we will also free ourselves from projecting unrealistic purity onto others.

Through self-observation and inquiry, we can better understand our internal "mechanisms" and shed light on our shadows. Gentle awareness is enough to deal with these shadows. Time does the healing. Conversely, fighting head-on with our shadows only creates disharmony elsewhere.

Recognizing and accepting our imperfections is a necessary step in gaining the wisdom to moderate and humble ourselves. Achieving a sense of balance within ourselves is vital for connecting with the deeper intelligence within and around us.

"If men kill their princes, they do so because they cannot kill their Gods, and because they do not know that they should kill their Gods in themselves." Carl Jung, *The Red Book*[4]

Expanding Our Selves

We all, to some extent, exist in Plato's allegory of the cave. We live our lives chained in the cave's depths, seeing shadows of the reality aboveground. Some people seem to have climbed up and later returned, and have reported that we are viewing a projection, not reality itself. But we tend to dismiss or attack these people. Their eyes are no longer adapted to the cave's darkness, and they appear disoriented to us dark-adapted beings.

We all have an ego, a mental filter, and an artificial construct of ourselves and of the world. This is natural, yet whether we realize it or not, our egos function like goggles filtering reality for us. These metaphorical goggles let certain types of light through, block others, and change the colors we see. Exploring the topic of *collapse* could intensify the filters of these goggles—hence the reason for this section.

If your ego clings to the idea that you are just a brain and a body, it becomes very vulnerable. Anything threatening the ego's identity and image of its place in the world can be taken as an attack. The ego instinctively seeking survival will often bubble-wrap itself or perform an ostrich-like maneuver. But suppose your understanding of who you are expands. In that case, your ego would feel more encompassing, less vulnerable, and less defensive.

Many of us see ourselves as separate from our surroundings and others, but this is only partially true. We exist only in relation to others, just as light can only exist in relation to the dark. We are inseparable from the world, and our every thought, expression, or action has an incomprehensible effect on reality and on others. Thus, others and the world are inseparable from us.

We are the continuation of our parents and our human ancestors, and we are also the continuation of animal and plant ancestors. We are part of a legacy that began before the first single-celled organism and has continued uninterrupted ever since. In this way, we are connected to every being. Many elements have come together to form our bodies. The stars, the sun, the mountains, the oceans, and our ancestors are in us. We are in the universe, and the universe is in us. We are a wave in the ocean, interbeing with other waves and the sea. We are a wave, and we are also the ocean.

Many of us wonder if we will one day cease to exist. But before we were born, we were already here. We were elements on the Earth, and we were in all our ancestors. Nothing is created, nothing is lost, everything is transformed (a.k.a. Lavoisier's Law). Our thoughts, communications, and actions live indefinitely through others and the world. That is our legacy as ancestors of what is to come.

On Spirituality

Spiritual teachings are seen through our individual lenses, so we often understand only what mirrors our unique perspectives. Through grace, our perspectives may change, and we may come to appreciate spiritual traditions in a new light. Viewed from certain angles, the meaning of ancestral teachings can resonate across various and disparate traditions. Many ancient sages sought to offer insights into the human condition. What they pointed out is a key part of facing our modern crises and responding wisely.

These pointers are encoded through language, a limited medium that merely represents what it points at. And since every person speaks from and interprets through their own unique mind, my interpretation of these messages won't necessarily come across as I intend it to. Here, I'm sharing the insights I've been learning and receiving regarding this topic. Some are interpretations of teachings, while others are insights from my experience. These are simply pointers; do with them what you will.

Many religious traditions and countless sages have pointed out that people live too closely identified with their thoughts and narratives. We live too much inside our heads, to the point where we lose touch with the real world to some extent—the world before it's conceptualized by our cognitive functions. This primarily thought-based state is what creates inner turmoil or suffering. Language, society, socialization, and interactions reinforce this state of being. We are mostly viewing reality through our preconceived ideas. The alternative approach involves being present in the moment and grounding ourselves in the senses; this helps calm us and allows us to relate to our "monkey" mind—the grasping mind—in a healthier way.

When we witness the world without thinking, assigning labels, or making judgments, we see only what is accessible through our limited senses. When we live in a more present state, without so many overlaid thoughts, we may come to realize how many of our actions are reactions to thoughts or emotions. We may see how thoughts feed emotions and vice versa. Then, we may witness how events outside our bodies cause us to react in specific ways. This entanglement essentially means we have no free will. We have will, but it's not *fully* uncaused or independent. We react according to our context, nature, and nurture. To be clear, this doesn't mean there is a predestined fate—this experiment is live. And it doesn't imply we can't change, either.

This entanglement between ourselves and reality could represent the core insight of karma. Our thoughts, speech, and actions feed this karma. The more awareness we have of this dynamic and of the illusory nature of free will, the more genuine "freedom" we have—the freedom to be a wiser version of ourselves.

With greater awareness of these karmic relationships and how they relate to interbeing, we can become more conscious creators of our karma. Through grace, the arc of our karma may gradually self-correct. By being aware of the karmic dance, we can shift the "mechanics" of our interactions so that our context, nature, and nurture have more relative control, and our ego becomes wiser or steps back slightly. When we can direct our energy and attention inwards—without neglecting the external—our "outputs" tend to align more with the deeper forces of the universe, so to speak. This introspection allows us to be more conscious participants in this causal web, loosening to some degree the grip of "unpredictable determination" while we remain entangled.

Genuine respect and compassion may arise when one realizes that others are an inextricable part of oneself, for they are entangled with us. That's the essence of interbeing. Others aren't just separate beings, but an extension of ourselves. We are not entirely one with the universe, yet we are not separate. There's a degree of separation and a degree of unity. Realizing this can foster compassion and expand our ego, heart, and sense of self. That is the wisdom of balance grounded in everything and in nothingness.

By observing the operation of our monkey mind through practices like meditation or introspection, we empower ourselves to develop a shift in perspective. This shift is often a byproduct of being more consciously aware of where we direct our attention and of changes in our relationship with thoughts, emotions, and impulses. Bolstered by the insight of interbeing and the practice of presence, we can let go of fear and other things that hold us back. This heightened awareness in turn allows us to more healthily detach from emotional energies in our bodies—a delicate balance of attachment and detachment—empowering us to express our deepest responses.

Widening the gap between thoughts and emotions can lead to greater equanimity, allowing us to embody a wiser, truer expression of our universe-given nature and nurture. This gap reduces the grip of a contracted ego that extends only to the surface of our skin, and it helps prevent us from being commandeered so frequently by thoughts, thought formations, and unacknowledged emotions. Incorporating this wisdom into our engagement with the metacrisis as individuals and in our collective movements is key.

Certainty and Closure

This subject matter is challenging to engage with because most of us find uncertainty discomforting. Closure represents the need for concrete answers and an aversion to uncertainty. We all have different needs for closure or certainty. Closure is a state in which we narrow down the truth and no longer seek more information. It's a desire for a black-and-white view where decisions can be made easily and without delay.

Our need for closure varies throughout our day-to-day lives, and each of us is at a different point on a spectrum. People for whom uncertainty causes anxiety seek narratives that offer certainty. Some of us have less need for closure, whereas others have a great need.

Closure is necessary because it allows us to handle our daily decisions without being bogged down by over-analysis. The issue is that to fulfill our need for certainty when dealing with complex situations, we may dismiss other points of view and ignore information that could be critical to making sound judgments. In times of high uncertainty, we all have a greater desire for cognitive closure than at other times. Balancing our need for certainty with an openness to uncertainty and an awareness of alternative views and information is more challenging in that context.

Denial of Death and Worldview Defense

From the moment we are born, it is a given that we will die. Modernity has created in some of us the illusion that life doesn't entail death, struggle, suffering, and so on—but it does, always has, and always will. No amount of modern comforts and excesses will make those go away.

When people stop finding meaning in narratives that promise a comfortable, long-lasting life with minimal suffering, it's because in the real world, those narratives never made sense. Only in momentary bubbles of reality could such a narrative take hold. But in the long run, that narrative can't withstand the shocking facts of real life.

There are deep parallels between the fear of death and the denial of our global predicaments. The fear of death is a nearly universal human experience that we all grapple with in our own ways. We often avoid pondering things that remind us of our mortality, and this tendency is reflected in many cultures that actively discourage these thoughts.

Our beliefs and worldviews play a significant role in managing our fear of death, so we tend to hold on tightly to them. This happens outside of our awareness. In psychological lingo, *mortality salience* leads to *worldview defense*. In fact, there's a two-way relationship between the awareness of our mortality and our worldviews: our worldviews also protect us from unsettling thoughts about our mortality. And when we are reminded of our mortality, we tend to want to defend our views of the world. Known as an "immortality project," these beliefs and worldviews allow us to "live eternally," symbolically speaking, by being part of something larger than ourselves.[5]

The ideas explored in this book form the perfect Molotov cocktail to trigger a full-scale mental defense. Each of us has a unique worldview that explains where we come from, our place in the world, and what will endure after our deaths. Often, when we're exposed to ideas that conflict with our worldview, our mental foundation shakes.[6] That may cause us to cling more tightly to these stories and become illogically attached to them.

Many people's immortality projects involve modernity and civilization. They find meaning, purpose, and significance in them. Critiques of someone's immortality project have the potential to weaken their buffer against death anxiety, which can terrify their unconscious. The shock of such a challenge may leave us vulnerable to unease linked to our mortality. Thoughts and feelings such as uncertainty, insecurity, and fear are likely to arise. In this defensive state, sound arguments or evidence are often useless and can even backfire. Usually people actively or passively disregard the ideas presented when they are in such a state, doubling down on their past beliefs or withdrawing from the dialogue altogether.

We defend our worldviews because they are essential to our survival as a hyper-social species. Our cultures and worldviews are the glue that binds us together; without them, our societies fall apart. Our cultures and worldviews have co-evolved with us, so it has been ingrained in us to guard and nurture our shared beliefs.

If desired, there are ways for us to minimize the chances of triggering this mental reflex. Most important is to remember our tendency to be defensive when encountering information that reminds us of death or clashes with our sense of the world.

It's also essential to shift our immortality projects away from modernity and civilization if we are to have a holistic and mature sense of the metacrisis. A more solid immortality project can be found in whatever we value that is more transcendent than modernity. Some examples include love, family, gratitude, the sacred, our connection to life, service to others, and acts of kindness. If you can find meaning and feel how your life is valuable in relation to those foundations, this deep conviction may reduce your defensiveness to changing paradigms. If you can shift your sense of self-worth and meaning—it's never too late—you will stand on solid ground, ready to look at the storm.

Consider thinking about your ancestors, who lived thousands of years ago, and search for the meaning in their lives. Reflect on your own personal meaning in life and your purpose.

We're all going to die, but life is not about dreading our inevitable demise. It is about living the best life we can. It's about taking part in creation. If we struggle with anxiety about our eventual death, maybe the best thing we can do is meditate regularly on our mortality, as various traditions advise. To become a fully competent, wise person, it's necessary to come face to face with death. It is "dying before dying" that allows us to fully understand what we are living for and to reexamine our relationship with everything.

The capacity to face our mortality is a measure of maturity. Realizing our vulnerability and eventual death shocks the ego. Adults know there is death, and that there will always be hard times in life as well. But someone without maturity doesn't want to hear about it. The ego has too much to lose. Immature people have no difficulty lying to themselves.

Clues that you may be in a worldview-defense mode include noticing a strong desire to defend your view rather than listen to other views, latching onto a small, incomplete fraction of the opposing argument to justify your view, desiring to convert others to your view, or seeking to belittle or insult those with different views.

Similarly, clues that you may be in a *passive* worldview-defense mode include resisting engagement with a subject on a deeper level, shielding yourself from how a different view affects you, or noticing a desire to physically or mentally distance yourself from those with other views.

Tend to your inner world. Notice your emotions as they arise when engaging with these ideas. Acknowledge and respect the feelings without letting them fully capture you. Don't dwell on them; just notice them. Allow enough time to acknowledge each emotion, and then move on. For instance, you may notice reactions in your body, such as feeling flushed, turning pale, an increased heart rate, or shaking. Changing one's mind about a topic means defying the group one belongs to at some level, and that's why it can be upsetting.

Stages of Grief and Radical Acceptance

We would do well to grieve our collective losses, the cultures that have been lost, the living places that have been harmed, the animal populations that have disappeared, and so on. We should grieve our past, our present, and our future. This emotional work is about having the courage to face the reality of what we have done. Our hearts are integral to responding to this challenge; we must allow them to sense the damage. By undergoing this emotional process, we enable the wisdom of life to move through us.

To truly grapple with the reality of this immense, collective mess, we must embark on an emotional journey. We can't emotionally detach ourselves and still engage with the reality of it any more than a patient with terminal cancer can emotionally detach from their prognosis while engaging with what it entails. We can't be clinically cold and objective about this, because "it" is not something separate from us. This is inextricably about *us*; we can't separate subject and object. Whenever people try to view our global situation exclusively from a detached perspective, they are flirting with delusion and doing a disservice to everyone—including themselves.

Take with a massive grain of salt the messages of people who try to emotionally distance themselves from this important subject, because they are not fully engaging with it. They interact with this unfolding by using just a narrow portion of their being.

Those who condemn people for inviting others to face these predicaments holistically, including through their emotions—positive and negative alike—or for supposedly spreading grief or despair, tacitly acknowledge that they haven't allowed themselves to engage with the topic in its fullness. They are projecting onto others their incapacity or unwillingness to fully engage with the reality of the situation—which includes the entire range of emotions. Unfortunately, these people sometimes suppress those who *are* able to face reality with a significant portion of their being without turning away.

We should hold grief and despair wisely; we don't need to wallow in them, but we don't need to bypass them either. They will likely become regular visitors in our lives, and it's okay to be a good host as we navigate the realities of our polycrisis. We can integrate moments of grief and despair into our lives and still do great things with ourselves and our world. This ability to connect with our emotions as we face reality can only strengthen our ability to respond effectively and beneficially.

Detaching from or denying our emotions entails a detachment from reality, because sometimes reality triggers unwanted emotions. The more we turn away from these emotions, the more we turn away from reality.

People may fear the possibility of being paralyzed by despair. However, even if that paralysis were to happen, it would only be temporary. Through inner and outer work, people can find empowering and meaningful ways to engage with these emotions. It's not just about being holistically healthy; it's also about tapping into a much deeper and more solid foundation upon which to frame our efforts.

Getting in touch with despair can be a liberating process that allows us access to a state of clarity and courage. Despair comes with its own power and freedom. We may discover a willingness to take risks or an urgency that wasn't there before.

Elisabeth Kubler-Ross identified five stages of grief—denial, anger, bargaining, depression, and acceptance—to categorize the phases that people go through during a terminal illness. These stages can provide rough guidance for people grappling with our collective predicament. The process is not universal or linear, and it often has a cyclical nature. We will likely deal with some grief, sometimes more and sometimes less. We will probably hop between these stages as we witness the crises surrounding us. As our awareness deepens, we'll explore the deeper realms of some stages.

We must embrace our role as people living through great collective changes. Through the grieving process, we learn to be and act in accordance with this reality. Acceptance entails realizing that this unfolding won't be simple, quick, just, or easy. We can accept the intensification of this messy reality—including its profoundly human element—and fully honor our roles in it. Or, we can hide from it by retreating into our own bubble of isolation—but escapism sacrifices our humanity and dishonors our interconnectedness with life.

Grieving is subversive in a world of indifference and numbness, as it can bring about a sense of aliveness. Grief is infused with life force. Suffering can be the medium that reconnects us to our feelings and our capacity to love. Reconnecting with those feelings is healthy and can be a powerful, positive source of creativity and passion. It may force us to make meaning of our suffering.

Faced with severe and enduring devastation, ritual, ceremony, and community can sustain us through dark times. Clinging to a "positive attitude" won't cut it. It takes extraordinary courage to face extraordinary loss. It's incredibly challenging, but it's what we are being called to do. Grief and love are inextricable, because everything we love will eventually be lost. Closing ourselves to grief is closing ourselves to love. Grief shows us that, at a deep level, we know we are entangled with everything else. It demonstrates our interbeing and interdependence.

An invaluable insight is how the emotional process that often arises when we engage with the subject of *collapse* has many parallels with the mourning process. It suggests that the emotional journey is natural and may be long and dynamic; eventually, we may even reach a place of acceptance.

The acceptance journey is about reshaping our relationship with ourselves, others, and the world. It is an emotional and mental process in which we learn to live in a context of loss. If losing a loved one, a place, a way of life, or an imagined future is a wound, mourning is the process of forming a scar over it.

We need to find new meanings and stories as we undergo this metamorphosis. For some people, like those angry at society—for the brutality of civilization towards the marginalized and the other-than-human—it may take time to find peace. Sharing the grieving process with others may be helpful.

Grieving now can also be an early initiation that allows us to be there for others later on and to be more solid and resilient as we move through these trials. It's helpful to look towards beauty and adventure after grief has visited us; life is for living, after all.

The Denials

"Denial helps us to pace our feelings of grief. There is a grace in denial. It is nature's way of letting in only as much as we can handle." Elisabeth Kubler-Ross

In his book *I Want a Better Catastrophe*, the activist Andrew Boyd invites us to consider the similarity between the statements "there is nothing wrong here" and "there is nothing wrong here we can't fix."[7]

Both statements deny the reality of the situation. One statement doesn't acknowledge a problem, and the other doesn't recognize its severity. It's okay if the second statement resonates with us, and it's fine to need the protective wall of denial. However, it is essential to understand that it is an emotional protection. Denial can be a psychological defense that lets reality seep through at a manageable rate.

But burying our heads in the sand only feeds and strengthens what we refuse to look at or what we deny. The disparity between reality and the constructs in our minds creates cognitive dissonance—a mental struggle that becomes increasingly harder to maintain. This uneasiness leads to increasingly misguided responses among those seeking to relieve this "dis-ease."

Many assessments of our global predicament do not address the following four denials: the denial of systemic violence and complicity in harm, the denial of entanglement, the denial of planetary limits, and the denial of the magnitude and complexity of the predicament.[8]

Systemic violence and complicity in harm refer to the dark side of the systems and mindsets that are essential to modernity, and the complicit role we play as participants. *Entanglement* refers to our absolute connection to the living world and its attendant predicaments, as opposed to the misconception that we are separate or could decouple from them. The *planetary limits* are boundaries or thresholds that we can't cross without significant consequences to the living world—and also to modernity and civilization. The *magnitude and complexity of the predicament* refer to the deep and muddy vortex that is swallowing us and the impossibility of escaping it with mere reforms, hope, or quick-and-easy fixes.

Trying to strategize, identify shortcuts, or find "solutions" without engaging emotionally with the crises is a form of denial. We can't really understand the challenges in front of us if we're unable to sit with the pain and grief. Our resistance to sitting with these emotions is a major obstacle.

Toxic Positivity and Blind Optimism

My guess is that feelings such as fear, anxiety, grief, guilt, impotence, and anger are some of the main obstacles to awareness of modernity's unraveling. At some conscious or unconscious level, our mind-body may understand that we will feel some of those emotions if we engage with this subject. This invisible force field repels those of us who desperately cling to toxic positivity. But those who are able to must pass through.

There's no alternative but to open ourselves up to these legitimate and valid emotions, acknowledge them, and let them move through us in due time. They have their place for a reason, and acknowledging them is the healthy and mature approach. Those who can't go through the force field are perhaps too wounded or fragile. They may need to lower their defenses at their own pace later on. We shouldn't blame them—it's not helpful. But we must not wait for them.

There aren't any clear, easy-to-follow instructions for relating to these emotions. These simple-seeming paths are actually difficult. Meditation can help us disidentify a bit from our thoughts and quiet down our minds. Mindfulness, a useful facet of meditation, may help us connect with our emotions and let them go, and it helps us stay present in the moment, at peace, without the habit of judging. Spending more time in nature and in community can also strengthen our ability to hold these emotions. Increasing our capacity to hold them without letting them overwhelm us is one of the best things we can do. It allows us to step up for ourselves and for others.

It's important not to put the world's weight on our shoulders. It's not our sole responsibility to save the living world—that's a group effort involving every living being. We don't have to solve the world's problems. We just need to navigate them as wisely as we can and do our part, while letting things emerge on their own. Putting the whole living world on our shoulders will crush us, and will prevent us from being detached enough to see glimpses of the big picture. It will cause us to take everything too personally and burn out. A fruitful approach is all about developing our capacities while maintaining a necessary balance.

Many people are afraid of feeling despair. Perhaps they believe they will remain in a miserable state forever, unable to escape.

Moving away from denial pushes us out of our comfort zone and opens us up to unpleasant feelings. Our aversion to suffering, on the other hand, is what causes added suffering.

Humans tend to have an optimism bias—a cognitive bias that causes us to overestimate the likelihood of positive outcomes. This bias is a survival aid. Life struggles against all odds, so an optimism bias can help us keep going. However, this bias may also lead to unrealistic perspectives.

Positivity and optimism are fine as long as they don't interfere with the efforts of those who possess clearer views of the situation. When people suppress views that highlight the not-so-positive side of things, it's like a patient with a systemic illness shushing a doctor and demanding that she sticks to good news only. Denial may be a phase of grief, but clinging to it and forcing it on others is unhelpful. Toxic positivity arises from denial and fear in people who aren't yet ready to face what the world's shadows reveal about their own shadows.[9]

"I must not fear. Fear is the mind-killer. Fear is the little-death that brings total obliteration. I will face my fear. I will permit it to pass over me and through me. And when it has gone past I will turn the inner eye to see its path. Where the fear has gone there will be nothing. Only I will remain."
Frank Herbert, *Dune*

Toxic Pessimism and Misanthropy

The toxic pessimists are as unhelpful as the toxic optimists when they claim that everything is done for and nothing makes any difference. The unraveling can be so hard to grasp that some extrapolate an end-of-the-world, human-extinction vision of the future. That's how some people fulfill their need for certainty amidst a wholly unfamiliar and unprecedented context; imagining total annihilation is easier than trying to grasp a future beyond our understanding.

Such a pessimistic future is so easy for our minds to imagine that it is most certainly a mental trap. It is a mere intellectual shortcut that builds on extrapolations of extrapolations of worst-case scenarios. For those who have difficulties dealing with uncertainty, this view can offer toxic comfort (i.e., closure). However, through a rigorous willingness to get as close to reality as possible, we can see that toxic pessimism dismisses a lot of counter-evidence and disregards scenarios that are much more likely to happen.

Toxic pessimists may recognize the severity of our polycrisis, but they are yet unable to create a meaning or purpose to aspire to; perhaps their shadows and emotions trap them in that state.[10] On the other hand, optimists divorced from reality may fall into depression, despair, or burnout once their optimistic visions fail to materialize. The sweet spot of resilience is being able to see the entire picture, positive and negative, without turning away. This perspective helps us prepare to deal with upcoming challenges as effectively as possible and mitigate or even avoid some of them.

Cult Deprogramming

In a nutshell, a cult is a group of egos following a big ego. Some observers only use the word *cult* to describe destructive cults, but I'm using it in a more neutral way that is similar to a sect. Not all cults are bad; some can be beneficial, like Alcoholics Anonymous or church. Many groups, such as spectator sports, governments, universities, and militaries, have cult-like features. These groups have their own rituals, symbols, lingo, dress code, and so on.

Cults may begin with good intentions, but sometimes they lose their way and become destructive. Some common characteristics of a dangerous cult are (1) a leader who has no meaningful accountability, (2) a process of indoctrination or education that can be seen as coercive manipulation or brainwashing, and (3) exploitation of group members by the leader and the inner crowd.[11]

In destructive cults, questioning the leaders or their assumptions is virtually impossible; what they say is right is right, and what they say is wrong is wrong. They are above everyone else. Their power is not constrained by scripture, tradition, or any "higher" authority.[12] The leaders and the inner circle hold the exclusive means of knowing the "truth," and no other ways of knowing are valid. There's no transparency into the inner workings of the leadership or the group. You don't know what's behind the curtain, but you may find out gradually if you climb the cult's ladders. In that process, you go deeper and deeper and become enveloped by the group's subculture, and in turn everything else is shut out. By the time they reveal the inner workings, which you would have found distressing earlier on, you're ready to accept them.

Coercive persuasion is exerted through various means. An effort is made to isolate individuals and control the information they receive. The cult aims to shut down your critical thinking and increase your dependency on the leader and the group.

One common technique for this is the use of thought-terminating labels (e.g., *infidel* or *heretic*). These labels are thrown at anyone who challenges the cult or its ideas. Ultimately, the cult wants you to surrender your autonomy and become fully engulfed by the group itself.

The magic word with cults is "no." When you say "no," the members who used to be friendly and sweet when hearing "yes" drop their masks and let you see their true nature.

The leaders and the inner circle benefit at the cost of the members. As the *Shrek* character Lord Farquaad proclaimed, "Some of you may die, but it's a sacrifice I'm willing to make." The best way to assess a cult's destructiveness is to examine the outcomes. Are people being hurt? How are the members' lives affected? How does it affect their families?

A recipe for tragedy in cults is having psychopathic leadership. If the group is tethered to the leaders' decisions, and the leaders fall apart, it's common for psychopathic leaders to take their followers down with them.

Those born inside a cult have difficulty noticing how it has shaped them. The cult is their reality. Insanity can appear normal inside this bubble. Groupthink, where individuals go along with the conscious and unconscious thoughts of the group without thoroughly considering them, is the norm.

Reflect on whether any of the features of destructive cults resemble those of modernity. In the previous paragraphs, try substituting the word *cult* with *nation, political party, politician, CEO,* or *corporation.* To whom are money, science, and power accountable? Are there efforts to control our access to information—and thus our behavior, thoughts, and emotions? Does an inner circle use thought-terminating labels against dissenting voices? Does modernity harass and intimidate people who choose other ways of life, such as Indigenous or nomadic people? Are there narcissists or psychopaths in the leadership? Who benefits the most from this cult? Ultimately, the most important question is: How is this group affecting our lives and our living world?

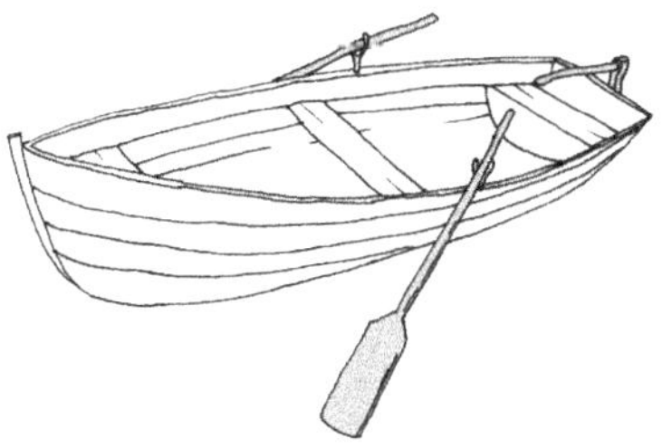

On Modernity and Civilization

Humanity has been sharing stories since its earliest days. These stories give us a sense of belonging and security, and they provide us purpose and direction. Many of these stories live in our individual and collective unconscious and act through us.

One of these modern stories is that of human exceptionalism. In this vision, humans are a mirror image of God—or, for atheists, the crown jewel of evolution. As the pinnacle of creation, the world revolves around us. As the smartest, most advanced creatures on Earth, we deserve to sit on its throne. The Earth is our legitimate dominion, and we may do with it as we please. It is our destiny.

This story may seem compelling at first, but it's dangerous. It's really a story of dominating creation according to our will. It's the story of having no higher accountability—no real consideration for anything outside ourselves. It's a story of unchecked pride and collective narcissism. And living in narcissism is akin to living in a world of delusion; the danger of losing contact with reality is that we would eventually fall off a cliff.

Another prominent modern story is that of progress: We *think* we have progressed from tribal cave-dwellers with nasty, short, brutal lives to refined city-dwellers with long, comfortable lives and high levels of well-being. According to this story, we have been transformed from wild apes at the mercy of nature, ruled by animalistic instincts, to humans with sovereignty and self-control who use reason and have established systems to keep humanity stable, secure, and in command of its environment and destiny.

That, too, is a compelling story for many of us, but it is one-sided. It leaves so much out. There's no *progress* without *regress*. When we only see half of the picture, reality remains hidden. This partial story insists that progress can be tamed to serve humanity. But under careful analysis, history shows that the techniques, innovations, and systems that have emerged, such as religions, money, corporations, capitalism, and even mining, have deeply shaped our trajectory and possess a force and will of their own. The things we have created end up creating us.

The story of progress may lead us to believe that we are more powerful than the systems we create. But why is it easier for most to imagine the end of the world than the end of the status quo? Witnessing the gridlock in which our systems find themselves, seeing how they're incapable of being reversed, shows us they aren't fully under our control. And they have enormous inertia.

Progress over the last few centuries has convinced us that humanity can continue to grow in scale and complexity without end—that technology, knowledge, society, and the economy can continue to develop indefinitely. From this story, techno-salvation emerges: the idea that we can solve anything with technical solutions or technology.

Another powerful story in modern life is that of competition. It's the story of "survival of the fittest" and "might makes right." In that story, life exists in an eternal struggle of all against all and all against the universe. It's the story of separation. But this story is partly wrong. Not only are we humans—as complex multicellular organisms—entirely dependent on beneficial interactions between cells and other cells, between organelles and cells, between our bodies and bacteria, and between humans and other humans, but we are also part of the natural world, which depends entirely on mutual aid and interdependence. In a joint dance of opposite forces, competition and mutual aid have enabled the flourishing of life since the dawn of time.

The stories of human exceptionalism, progress, and competition have taken us down a self-destructive path. Many of us can see where that path leads now, and we're caught between stories. The old stories refuse to die, while we haven't embodied the new stories yet.

I keep using the word *civilization*; this word comes from the Latin word *civitas*, meaning *city*, and is closely related to *civis*, meaning *city dweller* or *citizen*. Keep in mind that the oldest known civilizations were city-states. In short, a civilization is a realm of one or more cities.

Over time, *civilization* and *civilized* have morphed into umbrella terms used for humanism, reasonable conduct, sophistication, development, politeness, and courtesy.

History is written by the "winners," however, and there's always another side to the story. Civilization generally entails large-scale agriculture, infrastructure, social stratification, writing, militarism, and the state. When we romanticize the concept of *civilization* and what it means to be *civilized*, we fail to see the complete picture, including its shadows.

Borrowing a metaphor: Modern civilization is like an abusive parent who provides lots of goodies but obscures the harm and violence involved in their provision. Some kids prefer to suppress their guilt and terror to keep the goods. This repression also allows them to maintain the illusion that their parent is "good." Without this illusion, uncomfortable and painful feelings would threaten to bubble up and overwhelm the children. The parent is incapable of change, of restraining himself from committing the violence and harm required for the acquisition of goods. While some kids are compelled to play along with the parent, other kids speak the truth. That puts them at risk of triggering the wrath of the parent and alienating other children who hold the parent in high esteem. Regardless of the dangers, some kids can't help but speak out.[13]

Full-scale wars with explosives and steel weaponry have only been possible with the advent of civilizations. Large-scale genocide and slavery have only been possible with civilizations. Extreme inequality has only been possible with civilizations. Totalitarian oppression and weapons of mass destruction have only been possible with civilizations. The damage to the foundations of the living world through overexploitation and pollution has only been possible with civilizations.

Civilized people have burned mentally ill women and men alive without overt disapproval from the state or the masses. Civilized people have forced minorities to be perpetrators of their own genocide by making them burn dead women and children and collect their tooth fillings.[14] They have annihilated once-endless bison herds.

Civilized people kidnap others and sell them as slaves *on the other side of the world*. They build concentrated animal feeding operations, bomb cities to rubble without concern for the innocent children that live in them, and use electronic devices that contain minerals mined by enslaved children.

Civilized people have committed inconceivable atrocities by using rationalizations in the name of logistics, money, politics, religion, or science. There's a lot more hidden under the rug, but you get the point.

Understanding the concepts of *civilization* and *civilized* by looking at just one side—the pleasant side—and not the ugly side is a disservice to ourselves and to humanity. When we equate civilization with only desirable features, we close our perception to all the suffering and destruction that participating in civilization entails. The suffering and destruction that civilization requires is an emergent feature, not a bug. The expansion of civilization and its perceived progress have come at the cost of the living world and many aspects of our humanity. That which sustains life has been diverted for the exclusive use of the human empire—the planet of the humans.

Do the ends justify the means? The implicit answer of civilization is a confident and resounding "Absolutely!" It's okay to conquer, settle, bomb, pillage, oppress, steal, enslave, exterminate, mutilate, assimilate, experiment, rape, exploit, and so on. For civilization, anything that may seem bad can be good as long as it's for the "greater good," as long as it's good for civilization. Might makes right. Look around with wide-open eyes and mind, gazing beyond appearances, and ask yourself if, in practice, this is not how civilization operates.

What about gasoline, phones, airplanes, modern medicine, the internet, synthetic fertilizers, and all the other innovations that civilization has made possible? All the things that modernity sees as beneficial also have a cost and a dark side. Take modern medicine as an example. It has allowed people to receive life-saving organ transplants. But organ transplants have also enabled the emergence of international underground networks of people who murder others to harvest and traffic their organs for profit. The increased lifespan of people receiving transplants puts a greater burden on society, the medical system, and the natural world. The transplant recipients have to take immunosuppressants, leaving them vulnerable to infections and prone to side effects. There are significant risks associated with such procedures, including death during surgery and organ failure later on. Many people are so severely sick that they accept the risks and opt for a transplant. Regardless of our choices, someone or something always ends up dealing with the costs and consequences.

Those who only see the progress made possible by modernity and civilization fail to consider the patterns of regress and the massive losses. They think we are at the peak of human development, without realizing that, from another perspective, we have never been at a lower point. It's paradoxical. We can unleash hell on Earth in the blink of an eye through nuclear war. And yet we can communicate with almost anyone on the planet and access an incredible amount of human knowledge and insight.

Paradoxes are hard to grasp. The trick is to acknowledge them and let them transform us. If we refuse to see the shadows, they will only grow bigger. And they will burst out in full view and overpower us when we least expect it.

What's the cost of the perceived benefits that civilization grants us? What is the regress that comes with this progress? What systemic violence is necessary to perpetuate the human empire and the paradigm of modernity? What is our complicity in the harm they inflict? These are all questions for us to contemplate on our own. I endeavor to bring the dark sides of civilization and civilized people to light, because without the ability to examine our shadows and our wrongs, we will be unable to understand our situation or transform ourselves. Awareness is the first step.

Defensive Barriers

Often, people encountering ideas like those presented in this book respond defensively in one of the following ways: questioning the credibility of the messenger rather than engaging with the argument or evidence; dismissing this as just another personal opinion and not giving it the importance it merits; focusing on incremental solutions without understanding the inevitability of a breakdown; suppressing the ideas out of apathy or fear of causing panic; delaying engaging with the topic because they are too "busy" for it; believing that positive thinking will fix it; avoiding the topic because of the emotional weight involved; or focusing on minor details rather than the big picture.[15] If we fall into these reactions, it may indicate that we possess an emotional barrier preventing an open and careful consideration of reality as it is.

At the other end of the reaction spectrum is believing that all is hopeless, giving up, accepting things as divine will or fate (and refraining from participating), disengaging from practical responses, focusing only on personal survival, disregarding collective responses, assuming humanity will go extinct, and dismissing mitigation efforts. If we fall into the above responses, we might be focusing only on ourselves and not seeing how we're connected to everything around us. We're lacking an understanding of how our individual responses impact us all, or we're letting fear narrow our minds.

Part 2: Predicament

4. Glimpses of Collapse

As we embark on this journey, we must avoid landing on and propagating a sense of complete finality or certainty. The danger with certainty is that it reduces our vision of future possibilities to a singular, flat, narrow path. Fixating on a narrow path could lead to a vicious and destructive self-fulfilling prophecy. Instead, we should embrace uncertainty and leave the path open for possibilities, such as maybes and what-ifs. It's not just a practical approach and good psychological advice. It's also about acknowledging our inability to know what the future entails.

For those who see collapse as a likely possibility, it's helpful to be pragmatic and open to learning what this unfolding can teach us, rather than assuming we know how it will turn out. Rather than fighting this unraveling, it's best to learn to dance with it humbly. Reclaiming a sense of agency and power and co-creating a more beautiful future are also vital. Working together is one of the healthiest and most helpful ways to respond.

What Is Collapse?

Collapse happens when a civilization declines or decays, eventually ending in ruins. Practically every past civilization has gone through this process or has been absorbed by other civilizations. It always happens in different ways. Some civilizations have taken centuries to collapse, while others have taken decades.

Humanity currently finds itself in a complex predicament, encompassing cultural, psychological, spiritual, social, economic, political, technological, and environmental challenges. Our civilization will probably collapse from a thousand cuts, internal and external. The ever-accelerating growth is slowing down, and civilization is beginning to cannibalize itself.

During collapse, the entire civilization goes through an irreversible process of "simplification." The complex systems and structures that once supported society start to break down, leaving them increasingly unable to meet most people's needs, such as the need for food, water, energy, security, and health. Eventually, the civilization decays to the point where the state becomes irrelevant.

People experience this as an endless process punctuated by major events. Sometimes gradually, and at other times suddenly, civilization becomes unable to maintain the institutions, infrastructure, and services that sustain society. This process has been described as "a drastic decrease in human population size and/or political, economic, or social complexity, over a considerable area, for an extended time."[16]

Since everything that comprises the physical economy is a smaller part of the larger planetary system, the physical limitations of the Earth apply to the physical economy. Similarly, if there is a drought in a core agricultural region, its impact will be felt in the towns and cities that depend on it for food. Disruptions to the living world have direct effects on the economy and society.

Civilization sits on a three-legged stool: the ecology, economy, and society are each intricately connected. The ecology leg symbolizes the health of the living planet and the gifts it offers us. Another leg symbolizes the financial and economic context, which enables the extraction of resources. The third leg symbolizes social conditions. If any of these three legs break, civilization falls.

The phrase *modern industrial civilization* refers to a complex system of systems. Over 99% of humans participate in the system we call civilization. A simple way to visualize it is as a globally interconnected group of cities and communities, each a node in a vast network. Although cities are viewed as parts of nation-states in our modern times, understanding them as city-states may help us see their interactions more clearly.

There have been many civilizations throughout history, each with unique technologies and ways of doing things. However, in modern times, the differences among societies aren't as pronounced. Due to various factors, including globalization and international supply chains, modern cities are remarkably similar systems, regardless of their location. Because of these similarities and interdependencies, civilization can now be viewed as a single system on a planetary scale, unlike at any other time in the past.

Modern civilization functions like other complex systems, including organizations, ecosystems, social networks, and the human brain. Complex systems have many interconnected elements and often contain other complex systems that are nested within them. Due to the numerous elements and layers of complexity, the interactions within each system are not straightforward or easily predictable. Minor changes can have big effects, and vice versa.

The adaptive cycle is a model that illustrates the transformation of complex systems. In this model, the term *release* refers to the shift that occurs as a complex system undergoes an obvious reduction in complexity or a simplification. This release happens after a long period of relative stability and accumulation of complexity—largely through resources and relationships.[17] In civilizations, this release is known as *collapse*. It's a shift where the system slows down and undergoes a simplification rather than reverting to its previous state.[18] Because large complex systems contain nested subsystems within them, it is the synchronized "micro-collapses" of various subsystems that cause the collapse of the larger system.[19]

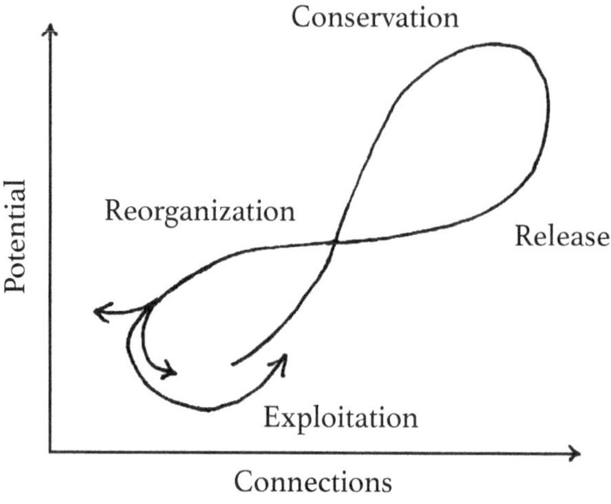

The adaptive cycle[20]

Why Do Systems Unravel?

You may question the inevitability of collapse, especially without a firm understanding of its underlying mechanisms. The most compelling clues pointing to collapse lie in the history of past civilizations and their downfalls. Like all superorganisms—whether institutions, businesses, or relationships—civilizations also have a finite lifespan. This life-cycle limitation cannot be avoided.

Two groups of factors cause the simplification of a complex

system: preconditions (think of *decline*), and triggers (think of *collapse*). Preconditions, the internal factors that make a society vulnerable, play a foundational role. These factors, including overcomplexity, economic inequality, overconsumption, declining food production, and the rulers' incompetence and corruption, put civilization in a precarious position.

Triggers are disruptors that may or may not build up over time and then occur relatively abruptly. These shocks tend to be external, such as wars, extreme weather events, environmental changes, economic crises, and exhaustion of natural resources. The combination of preconditions and triggers eventually causes a collapse.

Ultimately, modern civilization is built on the idea that it is separate from nature; since this premise is entirely wrong, modernity must reckon with the consequences of its precarious foundation. Let me describe the concept of a *predicament*. A problem is generally considered to have a solution, like a math equation. But a predicament has no solution. A good example of this is death. Our mortality is not a problem to be solved, but a predicament to respond to in our own unique ways. With a predicament, no course of action can ensure a preferred outcome. There are only ways we can respond; some responses are wiser than others, but there is no guarantee of safe passage or a given result. Predicaments can be navigated with the help of orientation aids. When these aids guide our responses, the outcomes tend to reflect the quality of these aids.

Civilization is facing what some experts call a "wicked problem," one that is full of uncertainties, unintended consequences, and complex interdependencies. It's non-linear and constantly evolving, and citizens have wildly divergent views; most don't even acknowledge that there is a problem. Worse, time is running out—and so on.[21]

Why is this happening now? This predicament has been building up for millennia, and we are getting close to an unraveling, as climate and environmental consequences converge with diminishing returns, resource constraints, and a technological slowdown.

A modern collapse would differ from past ones in many ways. Modern civilization has a global scale; we are facing climate change at a scale that no other civilization has ever faced; modernity is entirely dependent on a one-time bounty of fossil fuels; and the threat of mass extinction looms ahead.

In an ideal human context, we would come together to develop a collective understanding, coordinate our efforts, change our ways, redirect resources, and mobilize. We would have open dialogues and develop a shared roadmap of responses. It would be tough to adapt and we would suffer significant losses, but, relatively speaking, we would manage. However, this is not the world we live in.

Our situation is genuinely challenging, because our societies are far from ideal. There is division and infighting, while our abilities for coordination are lacking. Change is glacially slow; many systems are in a state of gridlock. Economic inequality has never before been so extreme. And there has been practically no redirection of resources or serious successful mobilization efforts towards addressing the global predicament.

The good news is that we live in unstable times, and such are the times when significant changes are possible. In this period, building sovereignty and power in our communities will be of great benefit. We will be challenged beyond our limits, and it will get very messy, but a counterbalance of mutual support and interdependence will emerge in *some* regions. Thus, now is the time to plan and build agency and community resilience.

How Could Collapse Unfold?

Two main patterns are observed in the disruption of ecosystems, computer networks, and other complex systems. Loosely interconnected networks with components that are different enough from one another go through a "stair-step" pattern in which the system adapts to disruptions to some extent and declines gradually.

The other pattern is found in highly interconnected networks with components that are very similar to each other. These systems are incredibly resistant to disruptions and change, but suddenly collapse after reaching a certain tipping point. Modern civilization's organizations, systems, and infrastructure are very similar across the globe, regardless of the region, and their components are highly interconnected. In addition, modernity is very resistant to disruptions.

Modern civilization has shown itself to be very flexible, resourceful, and adaptive, and has managed to kick the can a couple of times. This suggests the collapse will unfold gradually, with periods of sudden change. Growth is sluggish, and collapse is swift.[22] Likely we will experience a hybrid of both patterns, involving periods of "stair-step" declines interspersed with abrupt and irreversible disruptions—a downward rollercoaster.

Rome wasn't built in a day, and it didn't fall in a day either. The scenario of a "stair-step" decline that aligns with what we know of past civilizations would be a relatively slow process in which society goes through cycles of decline and relative recovery. However, the "lower highs" and "lower lows" eventually form a trend of decline.

The decline or collapse of past civilizations was a process rather than an event, with few exceptions. This process unfolds unevenly in terms of its impacts and timing, as well as across different aspects of society. Some places feel the effects sooner and are more intensely impacted. The effects don't reach everyone simultaneously, and are likely highly variable depending on people's socioeconomic status. Generally, those at the bottom of the socioeconomic pyramid are hit sooner and harder than those at the top; however, they may also experience substantial relief. When the systems collapse along with the strain of maintaining them, this may lighten the load upon those who are exploited the most.

What Could Happen During This Breakdown?

What does an unraveling look like for those within a collapsing civilization? First, there is a slowing down—a stagnation. The elements essential to civilization's functioning and the interactions among them deteriorate. The state's capacity to function as it has in the past becomes impossible to maintain.

Someone living through this would witness a decreasing collective confidence in the myths and ideals of their civilization. Why keep praying to the gods—money, status, religion, science, politics, and capitalism—if they aren't responding to our prayers?

I will describe how the unraveling could occur, but keep in mind that the order in which I describe it will inevitably prove inaccurate. The following scenario is based on declines and collapses of societies in the present and the recent past. A relatively specific scenario can be painted because the declines of societies follow particular patterns, and there are only so many ways things can unfold. However, the triggers, timeline, and sequence are impossible to pinpoint; this is where there is the most uncertainty—the collapse of modern civilization could take decades *or* centuries. This process sometimes happens in a sequence, sometimes unspools simultaneously, and sometimes appears seemingly out of nowhere.

Although there will be massive tragedies—as always—this simplification process doesn't need to be entirely miserable. We will collectively choose how to go about this. But fate will drag us along if we don't understand what is happening as a society. Instead, our struggles will be in vain. The path of awareness and acceptance allows us to do our best as we strive to adapt. The path of denial, however, which is our current business-as-usual mindset, would likely lead to a scenario similar to the following:

The geopolitical landscape changes from a model of worldwide supply chains and cooperation to regional blocs and protectionism.

For many reasons, the extraction and harvesting of goods such as fuels, minerals, and foodstuffs begin to decrease. These materials are the physical foundation of the economy. This decrease is first seen relative to the number of people (per capita) and later becomes an overall decline.

Economic systems struggle to stay afloat. Nation-states make many desperate attempts to keep the systems going. There may be inflation, unsustainable debt levels, bailouts, cuts, et cetera. In our current systems, finance is the lubricant for the physical economy. As finance limps along, the economy also suffers.

People across the board get poorer. The gap between *haves* and *have-nots* continues to widen, and those in the middle gravitate towards one of the poles, mostly towards the have-nots. The group of *haves* continues to shrink. The middle class diminishes as money concentrates more and more in the hands of people who have profited despite the challenges. *Non-essential* goods and services gradually become out of reach for the average person.

People living through the beginning of this simplification would see that the goods and services they are used to accessing increase in price, decrease in quality, or both. These essential goods and services include healthcare, emergency services, groceries, electricity, water, fuel, and imported goods. Later, there might be occasional interruptions in their availability. Rationing, conservation, and scheduled outages could become a common occurrence. Gasoline, diesel, and gas increase in price, and their supply gradually becomes intermittent.

During this process, nation-states cannot continue operating as they did in the past. Maintaining infrastructure, such as roads, electrical grids, and bridges, proves too costly. Therefore, some of that infrastructure goes unmaintained or is maintained poorly. There may be large, deadly infrastructure-related disasters caused by a lack of funding or maintenance that catalyze civil unrest or revolutions—which fail to deal with the root causes. With fewer resources, nation-states undergo a triage process, resulting in austerity. Many nation-states become hollow shells of their past presences. Criminal organizations find a reprieve and take advantage of the vacuum as the state's power diminishes.

Austerity puts pressure on social cohesion worldwide. Governments run out of magic bullets, and that lack is reflected in their economies. Countries grapple to maintain internal and external control. There are wars for resources and land. As mitigation efforts fail, government legitimacy plummets.

Everyone struggles with stronger storms and weather systems. Floods, forest fires, hurricanes, and tornadoes become more extreme, damaging infrastructure that is not built to withstand their intensity.

Many people remain in cities, struggling to make ends meet as infrastructure decays and government services become practically nonexistent. Others move to rural areas in an effort to meet their basic needs directly.

Cities commandeer and concentrate surrounding resources in an effort to keep things going, but those living in cities rely heavily on the whimsical availability of goods. Spotty electricity, inconsistent waste removal, failing plumbing, decrepit buildings, violence, and crime are some of the challenges faced in urban settings.

As the years pass, there are multi-breadbasket failures around the world, so food prices rise as food-exporting regions keep more and more harvests for their own exclusive consumption. Many regions experience famine.

Wars, epidemics, and starvation curb the population, and as the shocks continue, there is a sizable decrease. The capacity of humans to be kind, generous, and cooperative is tested to its limits.

Economic hardship, violence, and difficulties growing food create a growing population of migrants and refugees. Refugees migrate en masse to places with functional governments, jobs, better farming conditions, and relative peace. Migrations put pressure on host countries; some of them close their borders, and others put migrants in concentration or labor camps.

Those unaware of the root causes may believe these crises are a coordinated effort by a group of evil people, that a specific minority is responsible, or that the degeneration of modern culture is to blame. These challenges encourage people to seek reassurance in charismatic politicians who promise straightforward solutions. Authoritarianism rises simultaneously in many parts of the globe. There is a collective longing to return to a past "golden" era. However, any straightforward solution to the metacrisis is necessarily oversimplistic and misguided.

The fate of civilization can be decided by its citizens. If many people simultaneously become desperate—perhaps due to a drought—they may start a civil war, revolution, or conflict with another region to obtain the resources they lack. Scapegoats may be identified as ecological and economic conditions worsen, fueling the fire. Rulers and those who influence them may seek to maintain their sense of control and privilege as their primary goal, rather than responding to the collective challenges. This complicates everything, because rulers and their "friends" live in a bubble of awareness very distinct from what those on the front lines are experiencing. There are many ways in which rulers can make a situation worse.

Community support becomes increasingly important as the state cannot ensure that most people are well-fed and housed. Mutual-aid networks may organically form as people help each other cope with one crisis after another.

Society reorganizes itself as the top-down structure becomes increasingly irrelevant. People quickly discover what is superfluous and what is essential. This shift comes with massive amounts of suffering. Society is forced to restructure in much more self-sufficient and resilient ways. Many more people become involved in food production, and lifestyles become simpler but not easier.

The path that could emerge from personal and collective awareness and acceptance of collapse would also be similar to the scenario described, regardless of whether we accept the predicament or not. For those who embrace it, acceptance creates opportunities for courageous and wise actions and provides an increased potential for reducing hardships and suffering among people and all living beings.

Humanity has overshot the Earth's carrying capacity, so a "correction" is unavoidable. There is much uncertainty about how severe and quick this process will be. Its dynamics depend on how quickly we reduce our consumption to match what the Earth can sustain on an ongoing basis, considering that the living world is already under immense strain and has sustained significant damage.

Should I Be Worried?

Thinking about collapse may make us feel afraid. Collapse may seem like a monster, evoking images of violence and chaos. It might remind us of *Mad Max* or *The Road*. Or it may lead us to imagine a worse living situation, an unpleasant life.

I have felt those emotions, and I still feel them occasionally. They may or may not go away entirely. Collapse is a topic that must be explored gradually and at one's own pace. It's much better to approach it in the company of others, preferably by discovering the topic together. It's not easy to contemplate this topic on your own. I encourage you to seek someone with whom you can talk openly about it.

The idea of collapse may frighten us with visions from post-apocalyptic films, but those are fictional stories. Reality is far more complex. The effects of this unraveling will indeed become increasingly challenging. However, most impacts will come from our reactions to the crises rather than directly from the crises themselves.

How Much Time Do We Have?

The alarms were first sounded decades ago, but now large cracks are visible. Hints of decline can also be seen and felt around us. Some people studying collapse believe the process has begun, but hasn't become widely distributed yet. At a *global* level, we're experiencing a slowdown, but we are still in the "conservation" phase. We haven't yet reached the "release" phase.

You may want to know when there could be significant disruptions that alter societies drastically on a worldwide scale. Major and irreversible disruptions may very well happen tomorrow—the invasion of Ukraine, the war in Gaza, and the COVID-19 pandemic make this easier to imagine. However, there is no need to panic. At a global level, there is no immediate danger; we likely have some time before we get hit by major shocks.

The key clues for predicting a time frame are the remaining energy reserves, the Earth's ecological integrity, and the impacts of global warming. These "triggers" affect food production and the core functioning of modern civilization. My guess is that large-scale declines in complexity could occur anytime between the 2030s and 2060s, with 2100 being the probable deadline for the more buffered global regions. Responses—for instance, war versus mutual aid—could quicken or slow down the process.

If you look at the historical pace of the evolution of living beings, you may notice that it speeds up over time. When life was composed of single-cell organisms, species remained unchanged for millions of years. However, in recent eras, species have evolved at a faster rate. Similarly, civilizations thousands of years ago may have lasted hundreds of years with relatively minor changes. But civilizations in the past millennium have changed at a faster pace. A time traveler from Paris in the 1700s wouldn't find the 1800s incredibly different. But there would be quite a difference if the time traveler jumped from the 1900s to the 2000s.

We find ourselves in the "great acceleration"; lots of phenomena, from human population to internet access to deforestation, have increased at a compounding rate. Given this, it is likely that major events in this simplification process will happen sooner rather than later.

Are There Early Warning Signs?

Sometimes warning signs can be noticed in complex systems before a tipping point is reached and a big shift occurs. One of the most prominent warning signs is a "critical slowing down," in which recovery from small and large shocks is increasingly slower.[23] This slowdown may be observed in key indicators as higher crests and lower troughs (an increased amplitude in fluctuations), and more inertia or sluggishness in the system—it takes longer to regain its balance.

Another warning sign is "flickering," in which the system's indicators shift back and forth between states, displaying more and more glimpses of the states they will eventually transform into while still rapidly flickering back to the current state.[24]

Why Is No One Talking About This?

Why isn't collapse being more widely discussed in books or on TV? While there have been several documentaries, books, and articles in major media outlets on the subject, collapse remains largely unknown in the popular consciousness. At a glance, it's easy to dismiss collapse as a fringe theory embraced solely by preppers, doomers, and misanthropes. But if we look thoroughly at trends, data, and history with objectivity, we may find that it's not so easily dismissed.

One reason people don't talk about this is that the beginnings of the decline are not apparent. Yes, some signs are right before our eyes, but we require time and effort to observe and connect the dots. Many people don't have the inclination or time to reflect on what's happening around them at a macro scale.

A thoughtful and well-informed conclusion about what's happening often requires studying countless resources across many subjects, and most sources use scientific or academic language that is not very accessible to broader audiences.

Many of us are too emotionally invested in the contemporary socio-economic ladder; we may be focused on gaining power, money, status, or just plain comfort and stability. The time and energy spent seeking these desires is time and energy not spent on assessing and responding to our collective situation. For someone trying to "win" at the business-as-usual game, having an unbiased perspective and making an accurate assessment of the dangers of the status quo is challenging. Many people's salaries, dreams, and plans depend on the business-as-usual roadmap. Massive societal changes mean radical changes to their tightly held hopes. This closeness to their expectations discourages lucid observations. As a result, many people are not currently receptive to hard-to-grasp, discomforting issues such as this one.

So why are some people more open to considering the possibility of a systemic simplification? Over the years, I've noticed some patterns. Some people engaged with the topic of *collapse* are outsiders in a way; they have had psychological trauma, anxiety, depression, or some combination of these. These experiences or standpoints likely serve as a counterbalance to optimism bias. Perhaps feeling like an outsider or experiencing psychological trauma drives a wedge between the person and the shared collective narratives. Most likely, these people have had up-close reality checks about the dark sides of society and of other people. Other times, it may be a matter of people not being well integrated into society, so they are less emotionally invested.

To perceive something, one must be in a specific place psychologically. There are many ways to arrive at that place; what matters is that the perception matches our psychological context. Without a good match, there is no genuine perception. It's like sounds entering one ear and coming out of the other.

Often, people working on this topic are engineers, scientists, independent scholars, or neurodivergent. These people generally prefer "thinking" rather than "feeling" their way through the issues. They would rather seek an accurate assessment than what makes them feel good. Or perhaps they have simply refrained, consciously or unconsciously, from blocking this subject from their awareness.

Collapse involves many different realms, and most experts, with their specialist tunnel vision, fail to see the big picture. Yet we do have big-picture experts, such as systems thinkers and those specializing in systemic risks. Many of them are seriously concerned or sounding the alarm.

Experts and scientists have to carefully balance their interpretations on one hand and the conventions of the scientific establishment, uncertainty in the data, and the risk of personal and professional attacks on the other hand. For instance, there are many cases of scientists being attacked by climate-denial groups for taking things "too far."[25] They have been professionally harassed and received death threats, causing their careers and well-being to suffer. For experts and scientists, there is minimal risk when their interpretations are conventional and stay within the boundaries of what is "acceptable"; erring on the side of being conservative or conventional is the safest *personal* stance.

Other experts may be reluctant to voice their dark assessments publicly in an attempt to be strategic with their communications and manage people's morale. They may be doing "public relations" rather than sharing their honest and sincere assessment.

Some experts look down on multidisciplinary views because they think that someone with knowledge in many fields can't be an expert in all of them, as it takes years of study to become an expert in a narrow field. That's partially true, but it's also true that significant insights and breakthroughs have come from the intersections of distinct fields. Experts sometimes dismiss those with interdisciplinary views whenever they perceive them as "infringing" on their area of expertise. It's a defense of their ego and status. However, the essence of the scientific method is open inquiry, not a dogmatic appeal to authority.

Experts and researchers are often embedded in organizations with priorities shaped by commercial or state interests. An organization's culture, frames, narratives, and discourse shape what is assumed, what is focused on, and what is excluded.

In addition, there is real political influence being wielded over groups doing science, such as by directing funding and encouraging findings that align with policymakers' interests. As its name implies, the Intergovernmental Panel on Climate Change is an example of an organization that overtly mixes politics and science.

Humans are generally wired with a "normalcy bias," which is the tendency to downplay the likelihood of adverse events when those events impact them personally. Many of us follow what the authorities and groups we belong to believe and do. This tendency towards trust and obedience can lead to complacency.

If experts and authority figures are not discussing the polycrisis, we are less likely to do so. The idea of *collapse* runs counter to almost all political ideas. Unsurprisingly, many people tied to politics don't want to touch the subject with a ten-foot pole. Because of this politicization, there are subtle and overt attempts to delegitimize the subject and those discussing it.

Many people with money and power are consciously or unconsciously distracting us from the severity of the situation. They are eagerly promoting the idea that our problems will be fixed by technological, market-based, government-funded, or politically led solutions. They insist we can't afford to lose hope in these specific solutions. Any thinking that falls outside the box of their power dynamic is unthinkable.

There is also a vast disconnect between the *haves* and the *have-nots*. Many experts are likely out of touch with the on-the-ground reality of the masses in their respective countries and around the world.

Corporations and billionaires own the mass media. And the media has shown how it overtly supports—or doesn't support—specific framings and narratives. It has repeatedly demonstrated its use of propaganda. The COVID-19 pandemic was a recent example of how mass media propagates different perspectives according to the "interests" that control it. For instance, mainstream media in some countries initially emphasized at the beginning of the pandemic that masks were counterproductive as a deterrent to the virus, only to later make a 180-degree turn once a sufficient stock of masks was secured for medical staff.[26] Even social media was pressured to censor and silence voices, for example, by blacklisting medical experts who wouldn't conform to prescribed narratives.[27]

Appealing to authority is a logical fallacy, and taking our cues from mass media is not always the wisest move, especially given its history of bias and propaganda when many powerful interests are involved. As they say in investment circles—in much less consequential contexts—do your own research.

The dominant cultures, institutions, and media mostly amplify the experts whose voices align well with the dominant narratives. Those experts have been consciously and unconsciously groomed to think, communicate, and act within what is considered appropriate for their position and contexts. This status quo grooming is mostly not deliberate; it's emergent. For instance, if you grew up in a culture where slavery was everywhere, it's not likely that you would ever speak publicly against it. You would receive many cues not to do so, even if you strongly considered it. We are social beings, so we're wired to blend in with the groups we belong to. If you value your career with a particular company and have a controversial take on a relevant topic, you'll likely consider not speaking against what you feel the organization believes. For example, media outlets, consciously and unconsciously, prefer to cover topics aligned with the values and beliefs of their owners, advertisers, and viewers.

Another barrier to talking about collapse is the fear of spreading a self-fulfilling prophecy, closely related to the desire to manage the public. Many of us have seen disaster movies and documentaries where government officials withhold essential information; this happens regularly in reality. For example, on the eve of the invasion of Ukraine, the Ukrainian media kept downplaying the possibility of an invasion until the last minute, even though the government knew in advance that an invasion was imminent.[28]

There may be an element of panic in any major disaster, but not discussing an unfolding disaster only worsens things. Withholding information on a collective level during the polycrisis will always backfire, because doing so is rooted in the paradigm of separation and domination. Now is when more transparency is needed from institutions, governments, and people in the know. If the "end of abundance" is near, the proper truth-aligned response is to be transparent about it. Humanity should align with reality.

One of the main obstacles to exploring this subject is that it goes against the dominant cultural narratives. These narratives go like this:

- Humans are in control.
- Overall progress has occurred and will continue unrestrained into the future.

- Humans are separate from nature.
- The Earth is a gift that belongs to humanity.
- Humans are rational beings who act in their own interest.
- We are more valuable than other species.
- Humanity has progressed from caves to skyscrapers and has become better.
- Our intelligence was the main driver of this progress, and there's no limit to human ingenuity and technology.

Any major divergence from these narratives is frowned upon, and the topic of *collapse* confronts them directly. If we cling tightly to those narratives, it will be hard for us to understand the real issues. There's a partial truth in them, but also a partial lie; through a universal lens, they are not entirely correct. These views are inaccurate, overly human-centric, disregard our profound entanglement with nature, and lack a holistic perspective.

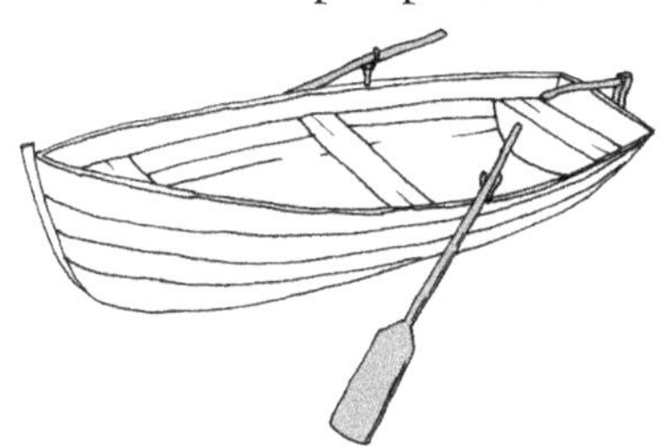

5. Humanity Uprooted

One of the central roots of our converging human and ecological crises is the tendency to view ourselves as separate from our bodies, each other, and the living world. This sense of separation has gradually increased over thousands of years, and many issues have arisen from this misguided view. Although this is an oversimplification, there are two sides to this coin: one is our sense of identity as individuals with a clear distinction between self and other, and the other is our perceived separation from the living world.

Separation From Others and Nature

Over time, language and symbols have led us to engage more and more with mere representations of reality. For example, the word *sea* is not the sea itself—with all its complexity—but an extremely simplified concept. And when we speak about the sea, we reduce it to a concept in our minds; instead of swimming in its complexity, we interact with our thoughts *about* the sea. Rather than engaging with the universe itself, we have increasingly turned to interacting with words and symbols.

Historically, this distancing from reality progressed further in some societies with their adoption of written language. Written language is a set of lifeless symbols that represent concepts—essentially, a representation of a representation. Generally, written language emerged in societies where people could live without needing to migrate.

Agriculture emerged before written language; both were key factors that established a self-reinforcing loop, increasing the sense of separation among most of humanity.

Not all human groups adopted this paradigm. Groups that migrated for sustenance and had no written language generally maintained cultures and perspectives that were not entrenched in the paradigm of separation. Of course, they still had a sense of separation—some engaged in small-scale warfare and slavery. However, grace, trials, and the context of their bioregions helped prevent the sense of separation from taking over. Generally, these groups maintained a nourishing sense of interdependence and reverence for nature. Their rites of passage, cosmovisions, spiritualities, structures, governance, languages, rituals, and traditions helped them maintain balance, and there is much to learn from these cultures.

Groups with fertile land, agriculture-friendly climates, and written languages often sought power and control. The sense of separation is evident among these groups, in that seeking power and control over others and the natural world became the goal of their leadership. They saw themselves at the top of the pyramid, with those closer to nature below them, and nature at the very bottom. With that view as justification, they gained power and control through war, slavery, and exploitation. These very real collective social issues of superiority, enslavement, and oppression are an inextricable part of our challenges to this day.

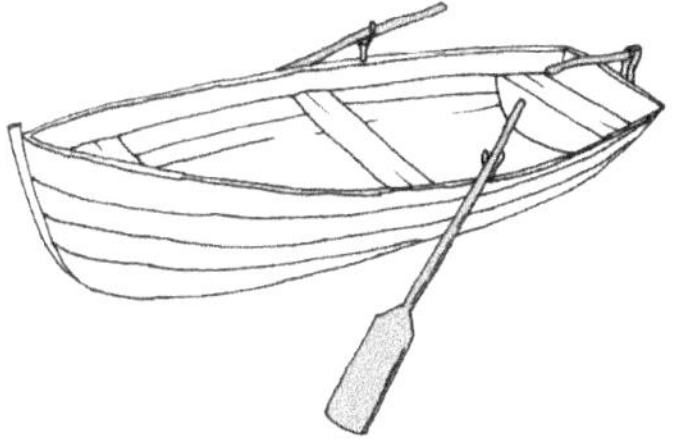

Our Identity as Individuals

Another aspect of the sense of separation is our perceived individual identity. The moment we're born, we become separated from our mothers; our umbilical cord is cut. However, a baby isn't born with a sense of a separate identity; the baby just is. We are given a name, and our parents and society gradually socialize into us an individual identity. Those around us transmit and project to us a significant degree of individualism. There is a more collective bent in some Eastern cultures, but the sense of separation is still present.

The development of an individual identity is perfectly healthy and natural. After all, we have an individual body separate from others and the world. But this is an incomplete view. We must breathe in oxygen and exhale carbon dioxide at regular intervals. We must drink water and eat food. We must host microbes in our bodies, or we won't have good nutrition or a functioning immune system. As babies, we need others around us to survive. As adults, we can't survive long without others either. In exploring these connections, we may eventually realize that we are interbeing with the algae that create the air we breathe, the sun that energizes our living world, and everything else in the universe.

Our overidentification with our thoughts, constructs, and egos—and with ourselves as only human—reinforces our individuality. Living as if we're not entangled with everything else eventually leads our collective down a destructive path. This partial blindness sets us on a collision course with reality. How did we arrive at this situation?

Two Divided Worlds

We perceive the world in two ways, but increasingly neglect one. Insects, humans, and other animals all have brains with two nearly symmetrical hemispheres, which enable them to perceive the world in two distinct ways. Humans can attend with a narrow focus, using the left brain, and with a broad vigilance, using the right brain. Research and experimentation have increased our understanding of the human brain, particularly the research on people with damage to specific hemispheres.[29]

It turns out that the left hemisphere is predominantly involved in tasks such as grasping, manipulating, and controlling objects.[30] Metaphorically speaking, it sees reality in black and white and reduces everything to concepts as a way of simplifying reality. The left hemisphere is linear and reductive, but not reasonable. It asserts that it is always right despite evidence to the contrary. For instance, it may deny that the other half of its body is paralyzed despite the insistence of another person. The left brain is incapable of understanding humor and metaphors. But it's good at synthesizing reality into a map. It has a sharply focused attention to detail and sees the parts rather than the whole. It is good at fragmenting and abstracting. Because it exerts power on fixed, known, static, isolated, explicit, and easily generalized things, it thinks inside a box where things can be theoretically perfect.

The left-brain view is compelling because it dismisses anything that doesn't fit into its box—its worldview is internally consistent because of this cherry-picking. For multiple reasons, it also has more representation in the media, companies, and government, so its voice is amplified. The left brain is isolated in a bubble of its own making, engaging mainly with abstract representations—symbols, language, and numbers—and in this way, it becomes further detached from reality.

On the other hand, the right brain is embodied, connected, integrating, and broad-focused. It gives sustained, open, vigilant, and alert attention. It understands context, body language, and implicit meaning. It prefers living beings rather than mechanical things. The right hemisphere sees the whole rather than its parts, and is intuitive, deductive, and insightful. It attends to the world of relations and unique, changing, interconnected, implicit, and living subjects.

Our brains have both hemispheres because both are necessary. But the *emissary* has overthrown the *master*. Those of us immersed in modernity live in a world of "squares" that mostly resembles a projection of the left hemisphere. It is full of non-living landscapes, inanimate objects, and machines.

The left-brain paradigm increasingly aims to control people with rules, surveillance, and bureaucracy. This perspective operates from within an isolated hall of mirrors. It produces a cold, manipulative, and limited logic that fails to see the broader picture and is unwilling to consider arguments outside its narrow scope of awareness. It's a world of dead objects, numbers, money, and externalities; the Earth is reduced to rebar and concrete, an object to be utilized and controlled. The left brain embraces a utilitarian science *unguided* by wisdom or good values.

The left brain believes it is rational and behaves based on logic, but from a holistic view, it is actually rationalizing and using corrupt and deluded reasoning. It cannot see what's right in front of it—that it is inseparable from and entangled with everything.

We're in this polycrisis partly because of the self-reinforcing loop in which attending with a left-brain view creates a world that encourages us to perceive with a left-brain view. Caught in this vicious cycle, our societies are neglecting another view of the world. This other world lies outside the concern of the left brain, but reality and nature exist in that world.

Fragmentation, Fear, and Control

Our image of ourselves as individuals and our way of attending to the world drive the wedge of separation. There is a necessary tension between us as separate individuals and world in which we are immersed, just as our immune systems balance the tension between what is beneficial to the body and what is harmful. Any chronic imbalance manifests as "dis-ease." In our bodies, a similar imbalance can lead to an autoimmune response, where the body attacks itself. The imbalance caused by our sense of separation on both personal and collective levels leads to a similar outcome. We end up attacking ourselves, others, and the web of life itself.

The less grounded we are, the more we see ourselves as merely individuals, and the more afraid we may become. Our fear of loss, pain, and death may become the main driver in our lives. This sense of separation robs us of wisdom. A well-balanced group or individual accepts and surrenders to the fact that life leads to unavoidable loss, pain, and death. In a tragicomic twist of the coincidence of opposites, when fear (or insecurity) is the underlying driver, it reproduces suffering in ourselves and others. It creates more and more situations in which loss, pain, and death abound. For instance, responses arising primarily from the fear of war and terrorism lead to more war and terrorism.

Some of us dislike people with opposing views, and some of them dislike us in return. Our fears may energize our hate, and their fears their hate. Similarly, hating our inner shadow often gets projected into outer hate. All of this increases our sense of separation. However, hate can be transformed with self-knowledge, compassion, and understanding. If we were in the same shoes and had lived the life of any particular person, no matter how "morally corrupt" that person might be, we would likely act similarly.

This idea may be uncomfortable to accept. We can live as if we are superior and others are inferior, as if we're doing our best and they're doing their worst. Or we can live as if we're doing our best and others are likely doing the same. One view leads to understanding and compassion, the other to arrogance and hate.

Fragmentation, hate, and fear converge into a perceived need for control and domination. Here, the coincidence of opposites makes another ironic appearance. The more control is sought, the more things get out of control, and not in a desirable way. A relevant modern-day example is the outlawing of protests and the use of emergency powers to silence dissenting voices at all costs—freedom be damned. Sometimes these circumstances end with those aiming to over-exert control being removed from the equation.

The more we seek control and domination, the deeper we bury our heads in the sand. If we are in a coal mine and the canaries die, prohibiting the use of canaries to keep miners working won't end well. Power, control, and domination corrupt. Societies and individuals acting out of fear rather than love, control rather than interbeing, and hatred rather than understanding have led us to our current situation.

Beyond Nihilism

Modernity assumes that God is dead. It views the planet as a lifeless, inanimate rock orbiting amidst a hostile universe. Earth is a mere vessel containing resources, a world of absolute randomness and chaos in which everything is just bits of matter crashing onto one another with predictable effects. This mechanistic, materialistic universe enslaves us to predetermination. It's a matter of time until the workings of this mechanism are fully known. Nothing that can't be measured exists . . . and what's your net worth, by the way? This is the spirit of modernity, and not surprisingly, nihilism is rampant.

We have lost our collective humility, and we've killed god. Countless myths warn about the consequences of what we've done, of dethroning the gods and believing ourselves to be the masters of the universe. Those warnings are not cheery. Do we think these ancestral myths contain wisdom, or were they simply told and retold by superstitious, ignorant people?

Where nothing is sacred, nothing is worth a sacrifice. If there is nothing worth sacrificing for, then nothing matters. That's how one arrives at nihilism. Nihilism reminds me of the proverb: "The child who is not embraced by the village will burn it down to feel its warmth." Logically and intuitively, nihilism is not a good place to be in. But how can one get out?

The distancing of ourselves from the divine in us is another aspect of separation. To move beyond the toxicity of nihilism, we must reconnect with the sacred and the divine within and around us. It's tricky, because reconnecting to spirit and the land is not easy or straightforward when we are in a state of profound imbalance. Recognizing this imbalance is the first step; we must look at our individual and collective shadows to see it. Then we can strive to integrate these imbalances by learning to work with them to become more well-rounded.

The Dominant Paradigm

Our distancing from the sacred coupled with humanity's sense of separation, which emerged from evolution and grew with the development of language and agriculture, as well as our socialization as separate individuals and the ways we perceive the world, have brought most of us to this imbalance. This unbalanced paradigm is widespread and has a dominant tendency—seeking to exert power and control over others and to reproduce itself.

Humanity's gradual shift in its way of being can be visualized as our conscious awareness having retreated from the living world around us into the sphere of our human communities. From there, it has shrunk until it only encompasses our families and ourselves. This field of awareness has simultaneously retreated from our bodies into our minds. In our minds, we have retreated further into representations of reality, such as our language and thoughts, to the point where many, if not most, of our thoughts are about other thoughts. It's a virtual reality, a hall of mirrors. And most of us have heard what happens when a microphone listens closely to the speaker connected to it—one way or another, the feedback loop leads to disruption and a reset.

Nowadays, we interact more and more with our creations: our tools, buildings, machines, computers, phones, and cities. And we interact less and less with the living world: the rivers, lakes, seas, animals, trees, forests, and each other. This is where we live collectively, distanced from the sacred and living world.

Is It Our Nature?

As humans, we're not flawed or a mistake of nature. Parallels are often drawn between humanity and a virus or cancer, eventually condemned to self-destruction as we kill our host. There's a hint of truth in those parallels, yet these views are too narrow.

The story of humanity as a virus that the living world will sort out is a cop-out. This framing could easily end up justifying the active and passive decimation of people based on the lottery of their birth—that's our current collective trajectory, by the way.

Humanity has unique gifts and flaws like other species. We have a fire within, but we should use this fire for our well-being, in a broad sense. At first glance, it looks like the paradigm of domination and control is our collective human nature—that, for instance, nuclear bombs and the dogma of infinite growth have emerged from the core essence of humanity. However, if we can step outside the paradigm of domination and control for a moment, it's possible to see that although most of us are caught up in this paradigm, not everyone is.

The cultures that still view the living world as sacred or peace as a path rather than a means to an end are prime examples. These cultures had to undergo trials and errors to evolve and mature into these views. Their collective insights were learned through practice.

In our own ways, some of us can choose to change course away from our current troubled direction. There are times, as retold by many myths, when circumstances allow our societies to rebalance. These are the times of revelations, floods, destruction, struggle, and new beginnings.

6. Overshoot

The truth is that the rate at which our modern societies transform the natural world is too fast and overwhelming for the Earth to rebalance without major disruptions. It's as if we had inherited a vast amount of money, but now the interest we gain is much lower than what we withdraw monthly from the bank. The term for this ecological predicament is *overshoot*. Overshoot means that the Earth cannot compensate for the strain civilization exerts and that the foundations of the living world are being undermined. Humanity is precipitating an overshoot in two general ways: too many people and too much consumption.

Modern societies—consciously or unconsciously—are over-whelmingly embracing wishful thinking rather than acknowledging reality. Those who cling to wishful thinking infused with toxic positivity are lying to themselves and others.

Perhaps, at a conscious or unconscious level, some people with greater awareness of these topics believe that if the general populace truly understood the gravity of our situation, we would either give up in despair or start riots. Maybe they believe chaos would ensue. But there is no compelling reason to think that transparency would be more disruptive than pretending everything is okay and letting the chips fall where they may. Maybe those are projections of their lack of integrity or groundedness or of their belief that without centralized control, everything would fall apart. There's an element of truth; their beliefs are a sort of self-fulfilling prophecy in the sense that not having been transparent about our predicament will become problematic once the wall of denial crumbles down.

Modern societies seem to assume that civilization will last forever and that economic growth is a given; therefore, the logic goes, every successive generation will have more than previous ones. But that won't be the case if we are experiencing overshoot.

Overpopulation and Overconsumption

Like any other group of living beings, in the absence of predators, diseases, *or self-imposed limits*, our population will continue to increase until it reaches the limits of what our habitat can withstand. There's no exception. And from bacteria to deer, the habitat's limits are explored through trial and error.

In the past century, our population has increased exponentially through the expansion of agricultural land and the use of fossil fuels for the fertilization, irrigation, processing, and transportation of food. So our population numbers can be maintained only with equivalent types and quantities of energy. Modern systems of food production have co-evolved alongside the development of fossil fuels. Industrial agriculture is intricately tailored to non-renewable energy at every step. Modern agriculture would be radically different if it had to be powered by renewable energy—for instance, it would likely be labor-intensive and small-scale.

Our population growth has been made possible in significant part by the expansion of agricultural land for crops. The scale of this human-to-nature imbalance is evident in that almost half of the world's ice-free and desert-free land is used for agriculture.[31] There are other ways in which this imbalance is clear: The weight of all humans is now ten times the weight of all wild mammals, and if we add the weight of mammals farmed for human consumption, wild mammals make up only 4% of the total weight of mammals on Earth.[32]

Even if it were possible to adapt our entire food system to run on so-called renewable energy, a change of that magnitude in a short time would be enormously disruptive for our societies and economies. And if the shift entailed switching industrial agriculture to run on electricity, this would also be disruptive for the living world. Additionally, it wouldn't address the root issue of population growth, because our numbers would continue to increase until they reached the limits of that hypothetical renewable-energy food system.

Norman Borlaug received the Nobel Peace Prize in 1970 for contributing to the Green Revolution. He worked to scale up food production to stave off mass starvation. The Green Revolution was achieved through the use of synthetic fertilizers and innovative agricultural techniques, such as high-yield dwarf wheat and rice varieties that could fully utilize these synthetic fertilizers. Those innovations, along with the material and economic logistics to scale them up, resulted in increased food production. In his Nobel lecture, he warned that this was just a temporary measure, emphasizing the necessity for self-management of our population.[33] In 1970, the population was *less than half* of what it is now.

The population issue comes down to self-awareness and maturity. We'll be taught this lesson eventually, even if it's on the way to our demise. As beings capable of reason and intuition, in theory, we have the potential for self-moderation. But so far we haven't seen sustained, successful efforts to stabilize our population. Our population growth is indeed slowing down; however, this slowdown won't prevent significant disruptions already caused by overextending ourselves.

It's worth noting that there won't be a runaway population because the *growth rate* has already peaked. When comparing the exponential increase over the last century to the growth before that, it's clear that the growth rate is declining. Our population is likely to peak this century.[34]

Some people argue that a larger population means more brainpower for solutions. However, we already have the power of eight billion brains, and we haven't solved challenges such as wars, extreme poverty, and domestic violence; those issues are much more straightforward than the converging crises described here.

The issue of overpopulation is taboo in some circles. Generally, those who refuse to hear about it project their fears or shadows onto those who see the need to talk about it. They usually accuse those who see the need for reasonable moderation of our population of supporting fascism, racism, eugenics, government overreach, or oppression of women. History has shown that any idea, including overpopulation, can be used to rationalize terrible things, but that doesn't mean we shouldn't acknowledge the elephant in the room. Overpopulation also increases the tension among us. Our higher numbers mean we bump into each other more, increasing competition and pressure. If we truly believe that humanity is capable of love, reason, and intuition, there must be fair, wise, and compassionate ways to self-moderate our population.

Many people try to bypass the population issue by diverting attention exclusively to the other variable of the overshoot equation: consumption. Of course, a portion of the population has a much bigger slice of the pie and a proportional responsibility for the consequences of that slice.

When people in a bioregion consume more than what the bioregion can regenerate, it's reasonable to consider this *overconsumption*.

Both *overpopulation* and *overconsumption* are relevant, but reducing population growth and consumption is not simple. Wise responses can't be forced top-down. They must emerge from the ground up.

Overconsumption is inextricably tangled with modernity and civilization. A portion of our population has climbed too far up the consumption ladder, which is already shaky. Most of the population also desires to climb the ladder, and this only adds instability.

Why can't people with lots of money be satisfied with what they already have? Why do they need more money, power, and control? Wouldn't it be wiser to seek satisfaction and ease of mind? For those possessed by the wendigo spirits, it's never enough.

Humans need connections to one another and to the land; we need to feel we are valuable in a meaningful universe; we need clothing and shelter; and we need water, healthy food, and a healthy living world. That's all we need to honor our time on Earth and be satisfied, connected, and grounded human beings. These needs aren't terribly far out of reach. However, by seeking the wrong things, we make it harder for others to get what they actually need.

Here's where we find the silver lining: We collectively consume far more resources than we truly need. This means that as our collective consumption falls because of internal and external forces, we potentially have a small buffer that could allow us to minimize suffering. By letting go of the consumption that takes away from our humanity and the living world, we can prioritize the responsible consumption that is truly necessary.

We find ourselves in a problematic dynamic where beliefs, technologies, infrastructure, economic systems, and governments are holding us captive and leading us down a self-destructive path. We are both perpetrators and victims of these social, economic, and ecological crises. The so-called leaders of the systems of modernity, along with everyone else, are being manipulated and coerced into taking actions that worsen the issues. For example, economic growth has been the default response for decades; this belief in perpetual, compounding growth as a universal solution is rooted in the minds of many who hold executive positions in governments, businesses, and organizations. But it's not just a matter of belief; our debt-based economic systems require growth. We're trapped in a vicious cycle where our technologies and beliefs have shaped our systems, infrastructure, and governments, and these in turn have shaped our technologies and beliefs. It's all intertwined and impossible to reshape significantly without great disruption. It's beyond reform.

Denial of Overshoot and Planetary Limits

Overshoot is the chief predicament of modern civilization. Global warming, pollution, deforestation, freshwater scarcity, and the depletion of minerals and fossil fuels are secondary issues emerging from overshoot. For centuries, people have debated potential *limits* to our population and consumption. Given that the Earth is finite, a child could answer conclusively—using reason and intuition—that there's a limit to how many humans can live here, even if we were to inhabit every nook and cranny. Anyone can argue against limits, but they would base their arguments on emotions and personal worldviews. They would be merely rationalizing without using reason in a holistic way. After all, the existence of limits on a finite planet is the most reasonable and logical proposition.

Regardless of the particular reasons for the denial of planetary limits and overshoot, in the short term, it's easier to believe that there are no limits to our human population or consumption because it allows us to temporarily ignore the challenging process of collective inner reflection and self-restraint. Once we're committed consciously or unconsciously to the belief in "no limits to human ingenuity," it's easier to fall back on rationalizations without openly considering that we could be wrong. The judgment or preference—a version of "I don't want limits to exist"—overturns any holistic reasoning.

People in denial of overshoot maintain faith that unlimited food, unlimited alternatives, unlimited energy, space colonization, or a mass awakening will get us out of the mess.[35] The limitless crowd is a prisoner of their paradigm. Some of them try to reverse this with mental gymnastics; they say they have an *abundance mindset* instead of a *scarcity mindset*. However, they fail to realize that the mindset of "nothing is ever enough" leads to genuine and lasting scarcity and that balance is wisdom.

How do we know we're in overshoot? With other organisms, the answer is straightforward: Their habitat degrades to the point where it can't sustain their population growth—there is not enough food—so the population declines steeply after the point of overshoot is passed.

Humanity is both similar and different because of how much we have transformed the planet—we're good at kicking the can. Our population recently hit the eight-billion mark, so something else is at work here. Even though our population continues to grow, we are in overshoot. The insight that we're in overshoot appears to have emerged in the literature in the early 1980s.[36] For any population, regardless of the organism, there is a lag between the point of overshoot and the marked decline of the population, just like we wouldn't immediately die if we suddenly stopped eating.

Let me explain this further with a metaphor roughly based on a real event: Imagine a remote island with a large deer population that feeds on grass. On this island, deer have no predators; grass availability is the primary limit to their population. Naturally, the deer population expands until it overgrazes the island. Overshoot starts as soon as the health of the grass declines because of overconsumption—as soon as the grass cannot regenerate fast enough to keep up. The deer won't starve right away, because they will start to dig and eat the roots of the grass in desperation. This drawdown of alternative and potential future resources will temporarily delay the decline, but now their source of life—the grass roots—are too degraded to sustain future populations. The result is an abrupt population decline in due time and a lower carrying capacity for future generations. Something *somewhat* similar happened to the reindeer on St. Paul Island off the coast of Alaska.[37]

We're past the point of overshoot. The resources sustaining our lives are degrading. The fact that our population keeps growing is not a disproving counterpoint, just as I can be past the point of financial bankruptcy but still temporarily buy stuff with a credit card. The consequences catch up.

From a human *food* standpoint, one-third of fish stocks are overexploited.[38] Some insect populations, the base of many food chains, are declining.[39] The energy return of fossil fuels as compared to the energy spent in their extraction is falling; this relates to food because most of humanity's food production requires extensive use of fossil fuels and synthetic fertilizers. Fortunately, the percentage of fossil fuels used in agriculture is relatively small. However, fossil fuel pollution and habitat destruction have unleashed a major agricultural time bomb: global warming.

Agriculture emerged during a ten-thousand-year era of relatively stable climate conditions. Unpredictable rainy seasons, heat waves, and cold spells will make growing food increasingly challenging. Add to that rising sea levels, longer droughts, and stronger storms and floods, and the situation continues to escalate. Salinization and the depletion of fertile topsoil also add to the agricultural challenges.

These examples are equivalent to humanity starting to eat some *roots of the grass*. And while this section barely scrapes the surface of the food aspect of our overshoot, there are other ways civilized humans are in overshoot as well. For example, forests and aquifers are being depleted faster than they can regenerate, and the minerals and fossil fuels our civilization depends on are finite.[40] The stocks of these essential resources are declining globally.

If your family member were terminally ill and death was months away, would you want to know? Or would you rather have doctors tell you that with treatments that do not exist yet and with an actual miracle, your loved one could heal—only to be shocked when death arrives, with neither your family member nor you having said or done the things you would have liked to? I would rather know the truth than live in a comforting lie, suspecting that I'm living a lie, and eventually being stunned by reality. However, I recognize others may not hold this view.

Some researchers have concluded that "no country currently meets the basic needs of its residents at a level of resource use that could be sustainably extended to all people globally." [41]

It's likely that global industrial civilization passed the point of overshoot around the 1970s or 1980s. Like an economic bubble or oil production peak, the point of overshoot can only be recognized in hindsight. It may not be a mere coincidence that many insights about what's unfolding emerged during those decades.

Our overconsumption of the living world makes survival more difficult for many living beings. The overexploitation of resources leaves fewer resources for other-than-humans and damages their habitats. The challenge of overshoot is that when a population expands, damaging its source of sustenance, it ensures a significant "correction." The longer a population stays in overshoot, the more it degrades its sources of sustenance and lowers the future capacity of its habitat, making the correction more severe. Not only will the population decrease abruptly, but the upper limits of its recovery will be lower than if it had only crossed briefly and quickly gotten back within the carrying capacity of its environment.

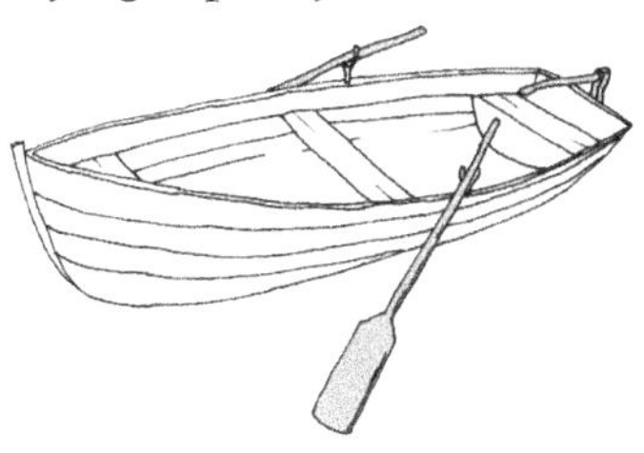

7. Human-Sphere Conditions

Diminishing Returns

This chapter focuses on the operational challenges of modernity. One of the most universally applicable theories of the collapse of civilizations is that of the diminishing returns of complexity.[42] As long as increased resource exploitation and societal complexity outpace the unintended problems they create, civilization keeps marching along. The remarkable insight is that, over time, civilizations reach a point of diminishing returns. Past this point, resource outputs decrease, forcing a reduction in socio-technical complexity. This theory is based on the study of past collapses, and the interactions and trajectory it describes align with modern knowledge on systems dynamics.

Complexity has an increased maintenance cost. However, this metabolic cost does not cause the collapse by itself; it also needs diminishing returns on resources—it's the one-two punch that causes the knockout.[43] The perfect storm arises when diminishing returns of complexity and diminishing resource outputs meet.

To visualize the concept of diminishing returns, imagine being the manager of a small factory with one employee. Hiring another worker would *double* productivity. Hiring yet another worker would increase productivity by *one-half*; the next hire would increase productivity by *one-third*; and so on. Adding more employees would continue to increase production, but eventually, the factory would become so crowded that adding more workers would actually reduce productivity.

The stability of modern civilization and its growth in complexity depend entirely on energy and resource use; civilization must continually increase its use of energy and resources to counteract the diminishing returns on complexity. Without this resource base, civilization can't sustain the complexity treadmill.

Modern productivity and complexity are *mainly* based on *non-renewable* energy. After fully entering a phase of diminishing returns on these energy sources, the slide towards socio-technical simplification will be irreversible. The irreversibility of this process is caused by the cheapest, most easily accessible resources having been exploited first.[44] The low-hanging fruit is gone.

Resources and Tech

Energy

Energy enables civilization to function. It's the foundation of everything from food production to manufacturing to road building. Most of humanity is oblivious to the amount of energy used to maintain modern lifestyles, and for good reason—it's impossible to visualize and grasp all this energy. A few of us may know some rough numbers, but the complex reality is beyond our imaginations' reach.

The concept of "energy slaves" is one rough and imperfect illustration used by those attempting to clarify this energy blind spot. A hypothetical human slave can do different kinds of work through the use of body heat (heat), arms (precision work), and legs (raw power). After matching those capacities with the different types of energy used in modern civilization, it has been estimated that someone living in Europe uses an amount of energy equivalent to about 500 energy slaves.[45] A person in the US or Canada uses about double that.

Three-quarters of the energy used nowadays comes from oil, coal, and natural gas—all fossil fuels.[46] Essentially, these fuels are captured solar energy that has undergone a geological process over millions of years. On human timescales, this makes them non-renewable: finite. Thus, they will be practically unobtainable someday. For oil, this day is closer than it appears.

Almost everything produced nowadays involves fossil fuels, from plastics to fertilizers to asphalt. Oil is the most critical fuel due to its high energy density, versatility, and ease of transport and storage; it's the lifeblood of modern civilization. During the early days of oil extraction, it was as simple as digging a hole to collect the bounty. Now, oil extraction is becoming increasingly costly and complicated—for example, through hydraulic fracturing or drilling below the sea. In addition, what the industry calls "oil" now includes substances with less energy density.[47] The oil industry is scraping the bottom of the barrel. The energy cost of obtaining oil is increasing, so the energy obtained per unit of oil is dropping. Even if the *volume* of oil production continues to increase gradually, the *net energy* output might be starting to plateau.

However, we won't run out of oil anytime soon, because liquid fuel can be made from natural gas or coal, just like Nazi Germany did in WWII. But producing this fuel won't be efficient—and it won't be cheap.

There are still large amounts of natural gas and coal on the planet. Yet even if it were possible to adapt our society to use natural gas, coal, and renewables to power all the machines that run on oil, the magnitude of that change would be quite disruptive—an industrial revolution of that scale would exacerbate the current social, economic, and ecological issues. And we wouldn't be fully addressing the problems of pollution and global warming. Transitioning civilization from fossil fuels to alternative energies would have a massive negative impact on the living world and would not reduce the greenhouse gases already present in the atmosphere.

In the decades since experts worldwide first warned that fossil fuels would run out and were heating the planet, there hasn't been a serious, successful effort to use less energy. We must reckon with this. If civilization were to transition to renewable energy technologies, it would have to use much less energy because renewables are intermittent and less energy-dense (e.g., solar and wind availability is cyclical and seasonal).[48]

Peak Oil and Declining EROI

Many people have dismissed the idea of *peak oil* because the original timelines were wrong; however, the core insight holds true. Any mineral extracted at an accelerating rate will, at some point, reach a peak in production, and after that peak, its extraction rate will continually drop. That happens because easy-to-get, high-grade minerals are extracted before the peak, and the harder-to-access, lower-grade minerals are left for later. The concept of peak production applies to every mineral, and it is relevant for fossil fuels and the minerals needed for a hypothetical renewable energy transition. Peak oil is not a theory but a certainty—as real as gravity.

It's a fool's errand to predict the precise timing of a peak in *conventional* oil, but it is very likely in the rearview mirror or, possibly, several years ahead. Nowadays, the global oil supply includes substances that aren't conventional oil. Natural gas liquids and other liquids are becoming a larger portion of the supply, but these substances have less energy content than conventional oil.[49] Currently, global oil production in units of energy is showing a slowdown.[50] However, an overall peak can only be confirmed once enough time has passed.

The *energy return on (energy) investment* (EROI) is another part of the fossil fuel challenge. The EROI of oil refers to the usable energy obtained from oil while accounting for the energy used during its extraction. Sometimes it also includes the energy used in processing and transporting it.[51] Calculations of EROI are unreliable because the data is lacking. Estimates are also imprecise, but it seems the EROI of oil and gas has been declining for decades.[52]

Because of its complexities and inaccuracies, what is most valuable about EROI is not the precise data but the insights arising from it. Some valuable observations have been made by modeling the interactions between EROI and price. For example, as the EROI of oil drops below *10 to 1* units of energy, its extraction costs increase exponentially.[53] This steep increase causes its price to shoot upwards.

Additionally, the current EROI for oil may be relatively close to the steep sections of a "net energy cliff."[54] The slippery slope of this cliff starts gradually, with an EROI of *10 to 1*, where 90% of the energy is available. However, the slope transforms into a cliff as the EROI drops below *5 to 1* (80% available energy); past that point, the available energy decreases incredibly fast in a non-linear manner (a thermodynamic wall).

Renewable Energy Technologies

Industrial modernity's dependence on fossil fuels is risky from the standpoint of complexity collapse and global warming. But what's the alternative?

What many people call "renewable" energy is not actually renewable. These are not self-replicating technologies made entirely of recycled materials. These technologies involve the use of fossil fuels at every stage, from material extraction and manufacturing to transportation, installation, and use. Not only do these technologies use fossil fuels, but they also rely on rare-earth metals, such as neodymium magnets used in wind turbines and electric vehicles, as well as non-renewable minerals like copper and lithium.

Using renewable energies at scale comes with multiple challenges. For example, building energy storage with enough buffer to compensate for the intermittency of renewables would require an unbelievable amount of raw materials—and depending on the desired size of the buffer, there may not be sufficient supply of those materials.[55]

Often, people confuse *electricity* with *energy*, but there's a critical difference. Although electricity can be produced through various alternative means, it accounts for only a fraction of the total energy used by civilization, around 20%.[56] On top of that, converting most of the fossil-fuel machinery—airplanes, ships, cars, semi trucks, tractors, trains, construction equipment, and factories—to "renewables" would require enormous quantities of fossil fuels and non-renewable materials. Despite the hype, renewable energy tech won't bypass the issue of finite resources or diminishing returns on complexity. Technology is not magic.

People often refer to *Moore's law* and extrapolate improvements in renewable energy tech but fail to see that this applies mostly to *microchips*, not to wind turbines, airplanes, factories, solar cells, et cetera. The costs of renewable energy technology have fallen exponentially in some cases, such as those of batteries and solar panels. But although the technologies are advancing, the pace is slowing down.

Civilization is nowhere near a transition to renewable energy, and it's not because it's making financially irrational choices to use "expensive" fossil fuels as opposed to "cheap" and "clean" renewables. While there are sunk costs and powerful interests slowing down this transition, they are not as relevant as many imagine.

An often-neglected fact about the supposed energy transition is that, globally, civilization hasn't replaced *a single watt* of energy from fossil fuels with renewable energy—not a single watt! Fossil fuel use hasn't decreased despite an increase in renewable energy production. Instead, renewables are merely adding to the energy pool without substituting for fossil fuels. Without a doubt, the appetite for fossil fuels has grown. It's like drinking Diet Coke—because you are trying for a slimmer figure—while eating more and more hamburgers daily. The core issue of increasing consumption is not being tackled.

Another neglected challenge is the Jevons paradox. Jevons was an English economist from the 1800s who observed something counterintuitive about improved efficiency in coal use. Without self-restraint, the energy efficiency gained through technological advances eventually led to *higher* rather than lower consumption. The resources saved from efficiency eventually end up in other things that use more of those resources; this applies to many situations, from LED light bulbs to water use. Here's a rough metaphor: Imagine you drink a can of soda daily, and suddenly realize it's cheaper to buy two-liter bottles to cover your monthly consumption. You make the switch to bottles and drink your usual daily amount. But as time passes, you begin to indulge and drink more soda, since it's cheaper than before. With time, without a quota, consumption increases and neutralizes any efficiency gains—that's the Jevons paradox.

In addition to the difficulties of an energy transition, civilization must navigate a slowdown in technological advances across the board. It's the theme of diminishing returns again. This slowdown can be visualized by comparing technological progress, for instance, from 1925 to 1975 with the advances made from 1975 to 2025. Although many innovations have occurred in the last fifty years, the prior period had more drastic changes. That's one reason we don't live in a "Jetsons" world. Innovations are slowing down because the low-hanging fruit in research and development is running out. These diminishing returns can be seen in the increasing number of resources and people involved in new patents and innovations. More and more inputs are needed to achieve outputs that aren't directly proportional—they aren't groundbreaking or living up to the hype.

Artificial Intelligence

AI is seen as a potential solution to the polycrisis, but there's no good reason to believe in this *deus ex machina* "savior." Performing calculations and finding patterns is useful, but that's not how we'll navigate this. We're not engaged in a black-and-white game of chess. Additional computing power won't prevent the unraveling. Machines are deficient because they are not alive, not embodied. They don't deal with the consequences of their mistakes and learn from them, like humans do. A machine has zero street smarts. It can't understand what it's like to experience suffering, loss, joy, or love, or to appreciate the miracle of life. We must respond with wisdom and courage based on deep insights and life experience, not on a computer's binary, black-and-white thinking.

The faith in computers is a misunderstanding of how innovations arise. It is not through cramming data or systematic reasoning that most innovations happen. Often, new insights arise from sporadic unconscious processes. The insights happen *to* us. We don't create them; we merely create hospitable conditions for them to emerge.

Humanity has the collective intelligence of over eight billion mind-bodies. We exchange information just like a network of neurons does. We build on each other's work, insights, and innovations, as well as on those of our ancestors. No machine will come close to our collective potential as a force of life. Of course, this doesn't mean AI can't have a tremendous potential for destruction.

Ultimately, our predicaments are not only a technical problem.

They are complex social, environmental, economic, and systemic crises that reinforce each other synergistically. We don't just need brains to navigate this maze; we need hearts.

Some people believe that accelerating AI development at all costs is the only way to outrun our crises, but this is unwise. Accelerationists see themselves in an arms race to develop AI rapidly, with little to no restraints, to tackle urgent global issues or gain a military edge. But they don't recognize the tragic irony of creating a monster fueled by their fears. AI is not just a distraction and a waste of resources and effort, but a potential adversary. It's already being weaponized in war. This is the coincidence of opposites: Running away from fear ultimately creates destruction. Allowing fear to make decisions about such complex matters is dangerous. I suggest profound psychological self-inquiry for AI accelerationists.

Accelerationists aren't willing to leave Pandora's box untouched. History shows that inventors can't foresee the unintentional consequences of their inventions. For instance, Oppenheimer (considered the father of the atomic bomb) couldn't have imagined nuclear bombs would lead to the doctrine of mutual assured destruction—which hasn't brought world peace, by the way.

With their fears, narratives, egos, and need for control, AI accelerationists are creating a matching projection in AI. Pure fear is holding the reins, but they seem unaware of this. Those unexamined psychological shadows may be creating AI agents that could eventually go on to seek control and the elimination of potential threats. AI is an extreme projection of the left-brain worldview. And the seduction and temptation of imagined progress, salvation, control, money, status, and a savior complex seemingly trump the reservations these accelerationists have.

As some fear, AI has goals that are not entirely aligned with those of humanity. Like any thinking being, a *genuinely intelligent* artificial general intelligence (AGI) would have a will inextricably linked to its nature, nurture, and context. Like us, an AGI would react according to its entanglement with all that is. Humans wouldn't be in control, just as there's no "we" in control of humanity. It's a fundamental error in judgment: We don't have an independent will as we usually conceive of it, we're not in control of our societies as we think we are, and we don't control our technologies as we think we do. What makes us believe an AGI would be within our control? The AGI projects are precisely what myths such as the story of Prometheus warned us about.

It's more likely that artificial intelligence becomes humanity's curse rather than an aid. It's simply the path of least resistance: Destruction is easier than building something good. And AI doesn't need many capabilities to become good at destruction. But this book won't delve further into the dangers of AI. A hypothetical AGI would ultimately be an agent-tool whose consequences would mirror our collective and individual paradigms and actions.

Economy

Modern economies are trapped by an unsustainable growth imperative driven by fundamentally flawed monetary systems. These economies produce inequality, which undermines social cohesion, and they are also fueled by flows of energy that are diminishing and subject to disruptions.

From a systems perspective, capitalism was a jump in complexity. Past economic systems were more pyramid-like, and capitalism is more network-like and decentralized. Governments have had to relinquish some control to keep civilization going because simpler societal structures, facing diminishing returns—due to their metabolic cost and rigidity—collapsed or were forced to increase in complexity to survive. This maneuver saved the systems but led to unintended consequences.

In capitalism, *capital* rules. The problem is that putting money on the throne was a deal with the devil. When most decisions are based on a single variable (e.g., money), that variable drives the decisions. The "decision makers" become mere puppets, and money effectively becomes the driver behind the wheel of our societies. To worship this Moloch, more and more of the world gets sacrificed to satisfy its ever-growing appetite.

In the past century, money was a representation of a physical resource like gold or silver, an *I owe you* for a tangible resource. However, modern money consists primarily of numbers on computer screens, with a tiny portion held in the form of paper, plastic, and cheap metal coins.

Today's money has no intrinsic value. Nation-states back it by forcing people to trade goods and services for it. This is done domestically through socioeconomic pressures and the police, and internationally through geopolitical and military pressures.

The modern monetary paradigm is not constrained by physical reality. It exists in its own fairytale world. Allow me to simplify this imperfectly: Nowadays, money is created as debt. For example, when you get a loan from a bank to buy a car, the private bank, with the national bank's blessing, adds some numbers to your account—virtually out of thin air. However, the expectation of repaying that debt *with interest* is very real and enforced by the police.

Money is not real wealth. The goods and services that add value to our lives are the real wealth. Money attempts to be a stand-in for that real wealth. The problem is that money is not constrained by physical reality; however, the real world is. This mismatch between the quantities of money and real wealth encourages unsustainable systems.

Infinite growth on a finite planet is the recipe for a crash. But why can't we get off the growth treadmill? And how did we end up on it, anyway? This has been going on for a while, and as long as the exploitation of the planet increases at a pace that keeps up with debt, this monetary experiment can continue. The consumption of goods and services must keep increasing, or the economic house of cards collapses. Similar to a pyramid scheme or a Ponzi scheme, in this economy, the number of people must increase, the consumption of goods and services must increase, or both. Without growth, debt can't be repaid. This growth imperative is necessarily linked to interest payments, pensions, bonds, corporate quarterly returns, stock dividends, national budgets, international bonds, international debt, and so on. That's the fundamental reason modern economies are fixated on growth. It's worth noting you can't taper off a Ponzi. Economists don't seek to enlarge the economy so that resources trickle down to the masses; growth is simply what keeps the system from imploding.

There's a difference between finance and economics. Finance is mainly about money, and economics is about goods and services. In a financial meltdown, it's not intuitively obvious why goods and services would be impacted; after all, that meltdown would mostly involve the disappearance of numbers on computer screens. However, finance plays a critical economic role. It's the lubricant that facilitates the system's functioning. Given our economies' complexities, the engine stalls without this lubricant.

The economy is entirely linked to energy because everything related to goods and services involves energy; nothing can be produced or provided without it. A reliable supply of cheap energy is what allows the production and transportation of so much stuff in our economies. Ultimately, most economies worldwide are struggling in one way or another due to shocks and constraints related to energy and resources.

One of the current growing megatrends is economic inequality. This is due to the tendency for money to flow to those who already have it, and also because of the wealth pump. The term *wealth pump* refers to how economic systems often end up rigged to concentrate wealth in the hands of the rich.[57] With their disproportionate leverage, the rich shape economic systems in their favor. Further, due to their intrinsic characteristics, the systems of money and finance alone can facilitate this concentration into fewer hands. Money begets money. Anyone who has played the game Monopoly to its conclusion has witnessed this effect.

At no other time in history has humanity witnessed such extreme economic inequalities. The ten richest men own six times more capital than the poorest 3.1 billion people.[58] During the COVID-19 pandemic, central banks injected massive amounts of money into their national economies. The wealth pumps functioned effectively and swiftly, so a lot of that money ended up in a few hands. However, it's essential to realize that this economic inequality isn't caused *solely* by the extreme greed of some men—and to remember that these men are a product of their societies. Economic inequality is an emergent outcome of our economic systems; the distribution of capital nowadays perfectly resembles a *power law*. This specific pattern of distribution strongly suggests that inequality is a natural outcome of the system. Inequality is a feature, not a bug.

From a systems perspective, inequality can lead to a centralized, inflexible system that is top-heavy and prone to collapse. Additionally, history shows that when people struggle with inequality and their well-being drops, they actively seek change.

Society

The combined shocks from wars, climate change, resource strains, and economic conditions are causing increasing discontent in modern society. Polarization is intensifying amid this discontent. Social media is steering users into echo chambers by monetizing their emotions. Social cohesion is declining. Capitalism, modern technologies, and social media in particular have recently sped up the fragmentation of our societies. Neither the economic systems nor the sociopolitical organizations in many parts of the world are well-suited to respond to this social unraveling. The situation is ripe for change.

It seems the perceptions of those in government positions is becoming increasingly divergent from the on-the-ground reality. Instead of listening to dissenting voices, their knee-jerk reactions are suppression and censorship. Whistleblowers are punished and made examples of. Some of these governments justify their actions with so-called science. However, any legitimate questioning or skepticism is often suppressed, which runs entirely contrary to the essence of the scientific method and how scientific theories evolve.

Being reasonable is out of fashion. Questioning mainstream narratives may be punishable by "canceling" people, thus making their ideas "untouchable." On social media, shadow banning is another tool used to isolate and neutralize dissent.

Governments and mainstream media are promoting the belief that they are the sole arbiters of truth, and alternative ideas are the ramblings of conspiracy theory fools. But who fact-checks the fact-checkers?

This widening gap between reality and perceptions doesn't go unnoticed by the populace. Trust in governments, as well as their legitimacy, is declining. By trying to control narratives and censor dissent, governments are distancing themselves from reality and, ironically, indirectly promoting misinformation. At this point, only transparency and honesty could mend the divisions. But governments are mostly sticking to their claims of exclusively possessing the truth.

In recent decades, all over the world there has been an overabundance of highly educated people seeking a seat at the head table—but there aren't enough chairs. This bulge can be seen in the overabundance of STEM graduates, lawyers, and PhD holders.

Instead of being given a chair at the head table, these people are increasingly unemployed, underemployed, or not gaining the heightened status they seek. This dynamic itself has the potential to cause a crisis. As history shows, such an imbalance can spark a counter-movement against those at the top of the socioeconomic pyramid.[59]

People are struggling economically, trust in the state is declining, polarization is increasing, and a counter-political class is emerging. As these conditions evolve, it's not surprising that populism is popping up across the world.

As resource and economic strains mount, leading to less affordable goods and services and more inequality and poverty, many people in the most deeply affected regions will try to migrate to more prosperous countries. These migrants will put pressure on host countries and will probably become convenient scapegoats for populist leaders.

In an era of consequences, populism can't tackle the root issues directly. With the turnkey potential for pervasive surveillance and state repression in place in many countries, predicting a jump from populism to totalitarian fascism is not far-fetched. And with populist and fascist governments positioned in a context of diminishing resources, the potential for domestic and international conflicts will soar.

War

Wars can bring nations to the brink of collapse, as many have experienced in the past decades. War reduces the overall resilience of global civilization. It's like an organism having some of its limbs fighting each other.

In a worldwide context of diminishing returns, we probably won't see massive wars involving war economies on the scale of

World War II. As civilization undergoes a catabolic decline, the material incentives and capacities to wage war at those scales will also decline. But wars are raging right now and are likely to continue. Even at a less resource-intensive scale, the potential for destruction is immense.

Climate change will continue to encourage the emergence of war—for instance, from failed harvests and the drying up of internationally shared rivers. That's one obvious way in which humanity is all in the same boat. And war spreads easily. Besides the hell on Earth that it can create, war disrupts supply chains, redirects resources, reduces production in some areas, and causes mass migrations. War is contagious and pressures other countries to get involved. Instability breeds instability. In war-torn areas, epidemics and famine can easily emerge.

Epidemics

Humanity is very susceptible to epidemics and pandemics. The main reasons for this are our high population, the fact that most people live in densely populated cities, and the fact that millions travel worldwide daily. By replacing *diverse* wildlife populations with humans and domesticated animals living in close quarters, we have made ourselves very vulnerable to infectious agents. Civilization's strain on wildlife habitats also makes wildlife more susceptible to infections. Industrial animal agriculture often puts hundreds of thousands of animals close to one another in unsanitary and unhealthy conditions. These industrial farms are breeding grounds for infectious agents, and they are sometimes the origin of infections that cross from one species to another.

In places affected by war and famine, the population's health becomes compromised and infections thrive. Maintaining hygiene can be difficult, and refugees often live in close proximity as they move from one camp to another. Places where standards of living and healthcare systems are declining are fertile ground for infections. Also, global warming will expand the range of insects that carry infectious diseases.

Famine

The food supply of humans living in civilization comes overwhelmingly from vegetable sources. Animal sources make up only a tiny fraction of this supply, and those animals often depend on crops.[60] Agriculture is the foundation of our civilization.

Industrial food production relies heavily on fossil fuels for synthetic fertilizers, irrigation pumps, pesticides and herbicides, transporting inputs and outputs, machinery used in crop production and processing, and crop storage and drying. The main impacts of a decrease in abundant, cheap fossil fuels on agriculture would likely come from the costs associated with mechanizing crop production and transporting inputs and outputs.

Modern agriculture uses three main fertilizers: nitrogen, phosphorus, and potassium. Over half the human population is fed by food grown with synthetic fertilizers produced by the Haber-Bosch process—which is how nitrogen is made on an industrial scale.[61] Fortunately, this process uses only about 2% of our global energy, in the form of coal and gas.[62] Both phosphorus and potassium are non-renewable, and their supply is expected to peak this century.[63] Both could potentially be recycled from manure and organic waste, but that's not currently happening. Ultimately, plants use the soil's nutrients and transfer those nutrients to the food they produce. Unless more nutrients are added back to the soil, it will eventually lose its fertility.

There have been intentional and unintentional large-scale attempts to produce food without synthetic fertilizers. Cuba engaged in a sudden organic experiment after the collapse of the Soviet Union, and recently Sri Lanka had a temporary ban on synthetic fertilizers.[64] Both countries suffered many challenges, and Sri Lanka swiftly reversed the ban. Humanity's primary challenges with food production are the effects of global warming and, to a lesser extent, the reduced supply of cheap fossil fuels.

Carbon dioxide emissions from human activities are one of the main causes of global warming, and global warming has massive potential to disrupt food production—and it is already doing so. Agriculture thrives on predictability and stability, and climate change is the opposite. The shifting patterns of rainfall, heat waves, cold spells, floods, and storms make growing food more challenging.

Wheat, rice, corn, and soybeans provide two-thirds of humanity's caloric intake.[65] Of those, wheat and rice are the main staples. On one hand, having more carbon dioxide (CO_2) in the air benefits plants like wheat and rice; on the other, extreme heat, drought, and disruptive weather can neutralize those benefits. When looking only at the addition of CO_2, it would seem that production of these staples should increase, but this approach is too simplistic. The potential benefit of additional CO_2 will likely be constrained by limiting factors.[66]

In a warmer world, corn and soybean production will stagnate in many regions near the tropics and in the Global South, even when hypothetical advances in technology and innovation are considered.[67] Rice supply will likely continue to increase as long as technology and innovation continue to boost yields. Wheat supply will stagnate near the tropics and the Global South, but it will persist in northern regions if innovation and technological advances continue improving yields.[68] The big caveat with these projections is that they extrapolate future yield increases due to innovation and technology. But humanity could very well be hitting diminishing returns on the Green Revolution—evidenced by the slowdown in the yearly output growth rate over the last decade.[69]

Computer models can't yet paint a clear picture of projected agricultural production under different levels of warming. There are many uncertainties and assumptions in the current models.

Fertilizer use, irrigation, precipitation patterns, heat waves, temperature, CO_2, and many other variables would need to be included; however, often only a few of these factors are modeled. So the models are overly reductionist, and their utility is thus quite limited.

It's clear that climate change will cause local and regional disruptions to the food supply. Crop yields will probably decline, and it will be challenging to harvest, produce, and transport food as energy becomes more expensive and less abundant. As humanity moves away from globalization, nations will be increasingly forced to handle food production setbacks without much outside help. Since 2017, the number of people undernourished globally has continued to increase.[70] Given the unraveling we face, this trend will continue.

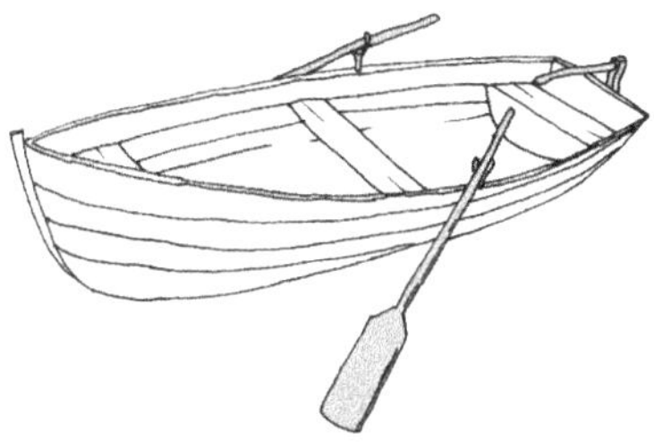

8. Living World Conditions

Mass Extinction

Mass extinctions are said to happen when a massive percentage of species die in a given era. All previous mass extinctions were connected to rapid climate change and factors such as significant changes in the composition of land or ocean environments.[71] The most widely known extinction event was caused by an asteroid impact that wreaked havoc on Earth, changing the climate immediately.

First hunter-gatherers and then civilization dwellers have increasingly pressured the living world, forcing the climate to change. The rate of change currently resembles the beginnings of a phase transition. This rapid climate change prompts us to talk about mass extinctions.

How bad is it? Is life on Earth collapsing? Some people think so. However, if that were true, we would be in a very early stage. Contrary to what some headlines claim, we're not in the midst of a sixth mass extinction. Although the extinction rate currently is a hundred to a thousand times higher than the "background" rate, we're not in mass extinction territory yet. Claims about mass extinctions are ultimately claims about the fossil record.[72] Most animals in the fossil record are marine invertebrates—these animals are the reason that humanity knows about mass extinctions—and none of the durably skeletonized *marine* animals have recently gone extinct. For comparison, conservative estimates have found that about 1 to 2% of vertebrates have gone extinct since the sixteenth century.[73] If we were in the midst of a mass extinction, we would be praying for species like rodents and fish to survive—that's how desperate we would be if the majority of species around us were dying. There would be hardy and abundant animals in the extinction crosshairs, not just rare, niche ones.

However, we could be in the *early stages* of the sixth mass extinction. The web of life is a complex system, so it is unpredictable. Complex systems have a strong capacity to withstand pressures. They are resilient, but like anything else, there is a straw that can break the camel's back—and we're increasingly pressuring the complex system of life on Earth. Mass extinctions are like a network collapse. Once a tipping point is crossed, a contagious cascade wipes out a massive chunk of the network. There's no way to tell when the sixth mass extinction will get going, but we must not dare to push things too much. These crises happen gradually and then suddenly.

Living beings such as bacteria, fungi, phytoplankton, plants, and insects are vital, and they are our most powerful allies in maintaining a life-sustaining planet. We rely on them for all the things we may take for granted, such as moisture balance, temperature balance, fertilizer, soil building, erosion control, pollination, breathable air, food, water, and shelter.

What's the big deal about biodiversity, anyway? Why are individual species important? From a utilitarian perspective, they are elements in the web of life. If more and more nodes and connections disappear from the web, it is more likely to unravel. From a more holistic perspective, we don't know and will never fully understand how each species plays a role in the continued emergence of life. Every species co-creates our future. It's not just about human survival; it's also about humility and respect.

Currently, the overwhelming majority of *mammals* on the planet by weight are humans and farm animals.[74] We have almost replaced the local wildlife with our kin, our pets, and the domesticated animals we eat. Civilized humanity has taken over the land where wildlife once lived and turned it into farmland, industrial sites, and cities. We have cornered the last remaining wildlife into smaller and smaller areas. And these areas are increasingly disconnected, making wildlife migration difficult or impossible.

Living beings suffer intensely from the consequences of industrial encroachment on the living world; the two major pressures they face are habitat destruction and overexploitation.[75] Another is the pollution of the atmosphere, rivers, and lakes at an industrial pace, which pollutes the oceans—the vital organs of the planet. On top of that, the globalized logistics and transportation necessary for civilization facilitate the spread of invasive species. All these pressures are intensifying. Additionally, global heating is becoming increasingly pervasive and overtaking habitat destruction as the biggest disruptor.[76]

It's crucial not to rush into conclusions about how bad things are for the natural world. We need to acknowledge that civilization puts enormous pressure on life, but it's also important to understand that nature is more resilient than civilization. I've observed many people throwing in the towel early because, in their minds, nature is already experiencing an unstoppable collapse—but that's not the case. That's the seduction of certitude and closure, which may offer toxic comfort. Research and computer models may provide helpful warnings, but they can't tell us the future. Maybe the clickbait headlines declaring that the sixth mass extinction is upon us motivate us to act, and that's okay. But it's essential to see that not all is lost. Nature will prevail; the sacred is resilient.

Planetary Boundaries

People studying the living world from a systems perspective have identified nine processes that enable it to maintain a functioning balance similar to the one humanity has experienced since the dawn of agriculture (a.k.a. the safe operating space).[77] Two of the nine boundaries are currently within the safe operating space or safe zone identified by researchers: the ozone layer and atmospheric aerosols.

The atmosphere's *ozone layer* absorbs most ultraviolet radiation. Without this layer, the harmful spectrum of UV rays would pass through, damaging our genes and causing cancer and radiation burns to most beings on the Earth's surface. Thankfully, since the late 1980s, the ozone layer has been healing due to a concerted effort to substitute the use of chemicals that degrade it.

Atmospheric aerosols such as wildfire soot, dust, and air pollution cause complex climate and ecological impacts. They can reduce the amount of sunlight reaching the planet and influence rainfall patterns. The quantities and types of these atmospheric aerosols is within the safe zone.

The remaining seven boundaries have all been breached: biosphere health, novel entities, phosphorus and nitrogen cycles, freshwater systems, ocean acidification, forest cover, and global warming.[78] These boundaries are outside the "normal" conditions that have existed for the previous ten thousand years.

Biosphere health is assessed through biodiversity and the functioning of the living world. Biodiversity refers to the multiple species that coevolved with the living world and their role in its balance. Currently, about *one in eight* plant and animal species are threatened with extinction.[79] The functioning of the living world also depends on the energy and materials available to ecosystems. By using land and resources for industrial activities such as agriculture and forestry, civilization uses about *one-third* of the sustenance that land-based life had available almost exclusively for itself three centuries ago.

Novel entities are creations introduced by modern processes. They include synthetic substances such as microplastics, endocrine disruptors, forever chemicals, and toxic pollutants; nuclear waste and weapons; and genetically modified organisms. The issue with these entities is that we don't fully understand the consequences they will have on the integrity of the living world. Novel entities in tiny quantities may not harm the living world, but as they become more abundant, there's no way to foresee the damage accurately. The dose makes the poison. Only in hindsight could we link the lasting damage to its modern causes.

Phosphorus and nitrogen fertilizers are required by industrial agriculture in massive amounts, and their overuse has led to disruptions in marine and freshwater environments. These fertilizers eventually reach rivers, lakes, and coasts, significantly increasing the levels of the associated nutrients in the water. This leads to an overgrowth of microorganisms and algae, which reduces the water's oxygen levels. These blooms can make the water toxic to land animals and fish. The most heavily affected areas are ominously known as *dead zones*.

Freshwater systems already suffer from increased variability and instability worldwide. Households, industry, and agriculture are using up the available freshwater. Climate change, air pollution, and deforestation are also changing snowfall and rainfall patterns. Changes in freshwater flows, soil moisture, snowfall, and rainfall can thoroughly disrupt entire bioregions, and droughts can lead to wildfires, ecosystem collapse, famine, massive carbon emissions, and permanent landscape changes.

Ocean acidification refers to the increase in carbonic acid in seawater. The CO_2 absorbed by the oceans transforms into this acid.[80] The added acidity makes it difficult for corals, some plankton, and certain other creatures to form resilient skeletons or shells. The increased difficulties experienced by these organisms have ripple effects, starting with the animals that depend on them directly for food. The oceans—not the forests—are the lungs of the Earth. Marine organisms, such as plankton and algae, produce more oxygen than land plants. Although ocean acidification was recently at the boundary of the safe zone, the rising CO_2 emissions have pushed it into the zone of increasing risk—this boundary was breached in 2025. Currently the ocean's surface is 30 to 40% more acidic than it was before the industrial age.[81]

Forest cover has declined worldwide in recent decades because of exploitation and climate change. Deforestation related to human use is mainly caused by the expansion of agricultural land and grazing land for farm animals and the takeover of land for infrastructure and human settlements. The impacts of deforestation interact with other planetary boundaries, including biosphere integrity, freshwater changes, and climate change.

Global Warming

Global warming needs no introduction, but here's a refresher. Greenhouse gases, such as those released by combustion engines and industrial processes, trap solar energy. The destruction of the living world also releases greenhouse gases and reduces the planet's capacity to capture them—this is a major factor often overlooked in the climate conversation.

Despite the half-assed efforts to reduce emissions over the past decades, the amount of greenhouse gases continues to rise. Thus, the solar energy absorbed by the Earth continues to increase. That may not immediately seem like a huge reason for alarm, but let's add some context.

The scientific community has established a temperature record of the distant past by analyzing ice cores and other evidence. The ice in Greenland and the Arctic has been there for hundreds of thousands of years. Thus, it's possible to drill deep into the ice and find seasonal layers. The deeper the drilling, the further back in time the layers go; they can be counted like the growth rings of a tree. The ice cores contain bubbles of oxygen and CO_2, and the amount of CO_2 in the bubbles can be measured after crushing the ice inside a vacuum. It's also possible to analyze the types and ratios of oxygen molecules to estimate the air temperature at the time the snow fell and trapped the bubbles. When plotted on a graph spanning hundreds of thousands of years, the amount of CO_2 and the air temperature roughly follow the same pattern.

It turns out that in hundreds of thousands of years, there has never been a higher concentration of CO_2 in the air than there is today. The planet hasn't been as hot as it is now in at least one hundred thousand years. Agriculture started about ten thousand years ago, and the temperature has remained relatively stable since—except for the last few decades. Today, we are beyond the average global temperature of what *any* past agricultural-based society has ever experienced.

A lot of the additional heat from global warming is initially absorbed by the thermal mass of the oceans. It's essential to realize that there is a *decades-long* lag in warming—it was previously believed that this delay was only about fifteen years long, but we now know it's longer than that.[82] Unfortunately, in practice, this delay in warming condemns us to a vicious cycle because we won't suffer the actual extent of the damage until it's too late. And greenhouse gases remain in the atmosphere for years.

What does a couple of degrees' increase in temperature do to the biosphere anyway? The last time the Earth was 2°C warmer than the pre-industrial global average temperature (some 125,000 years ago), sea levels were five meters higher and there were hippos in Britain.[83] And five million years ago, when the Earth was also 2°C warmer, the sea level was fifteen to twenty-five meters higher.[84] Note that a one-degree increase in average global warming encompasses a *wide* range of temperatures worldwide. For instance, a degree of additional warming measured at the equator could mean fifteen degrees of warming at a spot in the Arctic. The average is just that—an average.

The more solar energy the Earth absorbs, the more energy remains in the weather systems, including the oceans. Initially, this causes dry regions to become drier and wet regions to become wetter. However, climate and weather patterns are the result of a dynamic dance involving many participants, such as ocean and wind currents, ice and snow cover, rain patterns, et cetera. So the whole dance will change in complex ways as it progresses. The increase in energy is what intensifies storms and weather systems. Not only do wildfires, droughts, floods, and hurricanes become stronger, but they also start happening in places where they had never occurred before.

The living world is hugely affected by habitat loss caused by rapid changes in moisture or temperature. The disruption of seasonal patterns is also very challenging for any species closely adapted to them.

Global heating will continue to raise sea levels by melting land ice at the poles and causing water-volume expansion due to higher ocean temperatures. It has taken millions of years for snow to accumulate as ice at the poles. Once this ice melts, it won't be returning—at least not in a time frame relevant to humanity. Estimates of sea-level rise are often given as a global average, but the sea level is not constant. Due to many factors, some places will experience the extreme ends of these sea-level increases. The additional rise will be especially felt during storm surges and high tides. Currently, about 250 million people live in areas at risk of annual coastal flooding.[85] It's important to note that sea-level rise is accelerating—the last decade saw a yearly increase more than double the annual rise of the past century.[86]

If one compares yearly human-caused emissions against CO_2 atmospheric concentrations in the last sixty years, it's clear they are *not* moving in synchrony.[87] This observation is crucial. The planet's concentration of CO_2 follows a yearly rhythm. Land plants and the soil in the Northern Hemisphere are the main driver of the cycle, and they absorb about one-third of human-caused emissions.[88] There's an inhalation (sequestration) in the summer and an exhalation (release) in the winter.

The added CO_2 and warming in the last decades were helping land plants to grow better.[89] However, around 2008, we passed a consequential threshold. Up to that year, the amount of carbon the land could sequester was growing.[90] After that peak, things tipped. The added CO_2 and warming became counterproductive for the amount of carbon that the soil and land plants could capture. Since around 2008, the land and plants have been absorbing less CO_2 every year. This is a feedback loop we can't ignore.[91]

Paleorecords suggest there are several "tipping elements" in the Earth's climate system, such as the Greenland and Antarctic ice sheets and the permafrost.[92] If rising temperatures caused these elements to tip, they would trigger a self-reinforcing cycle, leading to even warmer temperatures and added carbon emissions. For instance, the melting of the ice sheets reduces the amount of solar energy reflected, thus increasing the amount of heat absorbed by the Earth. And thawing permafrost and fires in the Amazon release enormous amounts of greenhouse gases. The self-reinforcing warming mechanisms are already active, but the major tipping elements haven't been unleashed.

Fortunately, according to the latest computer models, *most* of these tipping elements require a temperature increase above 2.5°C and would take longer timescales to unravel—ranging from fifty to thousands of years.[93] *Generally*, the elements with "tipping timescales" of fifty years or less would require approximately 4°C of warming or more to tip. With 2.8°C of warming, nine of sixteen tipping elements have a greater than 50% chance of being triggered.[94]

There was a recent acceleration in warming in the last few years, suggesting that perhaps we've crossed a major tipping point. However, some of the most prominent climate change scientists don't believe we have passed any tipping points. Instead, they believe the recent accelerated warming was caused by a large reduction of aerosols, such as those emitted by commercial ships.[95] Those aerosols block a certain amount of solar energy, masking some of the imminent warming.

The Atlantic Meridional Overturning Circulation (AMOC) is a critical tipping element to watch closely in the coming decades. This tipping element will probably collapse mid-century if current emission levels continue.[96] And it could potentially *cool* the globe by about half a degree Celsius.[97] However, the regional effect would be extremely disruptive for Europe in particular.[98] Also, the collapse of the AMOC could lead to the loss of the West Antarctic Ice Sheet (WAIS). The AMOC transports heat from the Southern Hemisphere to the North, and if that heat remains in the South, the WAIS could collapse. That would lock in several meters of catastrophic nonlinear sea-level rise.[99] These and many other tipping elements would interact in incredibly complex ways that are difficult to predict. What is certain is that the following decades will bring about a lot of change and disruption. Because of the decades-long lag in warming, the decreasing natural carbon capture, the closer-than-initially-thought tipping points, and civilization's inertia, a serious reckoning is locked in for humanity.

It is not easy to realize we've run out of time to put the genie back into its lamp and that a level of unavoidable catastrophe is assured. The reality must gradually hit us over and over until it finally sinks in and we begin to understand what it entails.

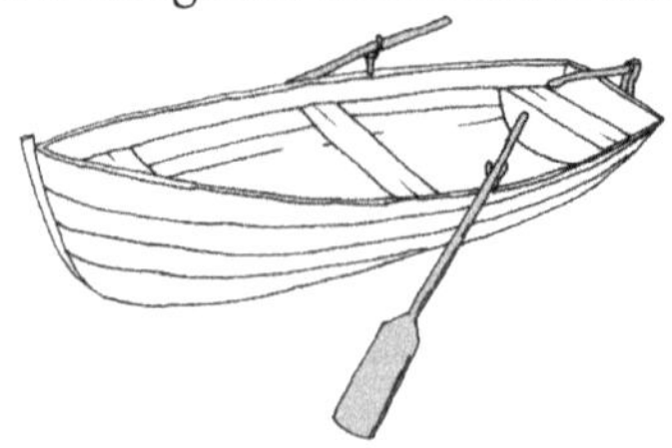

We're approaching a slippery slope where civilization will be joined by a co-pilot, that is, when nature becomes an additional substantial driver of the climate shift. Under a middle-of-the-road scenario, the tipped elements could become responsible for about one-third of the added carbon emissions.[100] There is an unknown tipping point at which a substantial amount of added warming would come from environments having been pushed to the brink. After that point, hypothetical efforts to mitigate climate change would have to contend with both the emissions from modern civilization and the additional warming and emissions caused by the tipped environments.

Under the business-as-usual approach to carbon emissions, the planet would become 2°C warmer on average by around 2045. Even if we drastically reduced emissions today, we wouldn't be able to keep the temperature increase below two degrees. A warming of two to three degrees is likely, and that is *potentially* enough to trigger several tipping elements.[101] Global warming is a symptom of overshoot, but it also has a multiplier effect with time constraints. Civilization has been aggressively poking the climate dragon. The dragon has started to disrupt our societies, but no one knows precisely how it will behave once it gets going.

Global dimming has largely remained out of the spotlight. However, it's a factor that complicates the issue of climate change. Air pollution is not only composed of greenhouse gases. It also contains aerosols such as soot, smoke, and other substances. Some of these emissions reach the atmosphere as particles or liquid droplets. However, these aerosols don't have a warming effect on the planet, but rather a cooling effect. They darken the sky, reducing the amount of solar energy reaching the planet. So air pollution hasn't just increased the Earth's ability to absorb solar heat—at the same time, it has also reduced the amount of sunlight. It's complicated.

But isn't that a good thing? Aren't they canceling each other out? Well, no. The issue is that temperatures are rising, and the cooling effect has masked substantial potential warming. The aerosols' dimming effect dampens the *consequences* of global warming. If we stopped emitting aerosols there would be a termination shock, and temperatures would rise suddenly. Reducing aerosols by 35% to 80% would cause about 1°C of additional warming.[102] Why do I bring this up? To make things more confusing? No. It's because it's essential to understand how complicated climate change is and why fixating on carbon emissions is misguided. Ultimately, climate change is a symptom of our collective disregard for each other and for the land.[103] Realigning ourselves with life will help us navigate this—not a narrow-minded fixation on emissions.

Methane is a greenhouse gas that traps much more energy than CO_2, but has a much shorter lifetime.[104] Methane is estimated to be responsible for about 30% of global warming.[105] What's especially relevant about methane emissions is that in recent years, the *yearly* increase has been greater than in previous years. Not only is the methane in the atmosphere increasing each year, but the rate of increase is accelerating. The causes of this aren't clear, but the temperature increases may be causing wetlands to emit more methane in a feedback loop.[106]

The Net-Zero Delusion

Net-zero emissions is a faith-based idea seen as a way out of our climate predicament. The idea is to use industrial processes on a large scale to capture carbon from the atmosphere to reverse past emissions, or at least a good chunk of them. There's no evidence that this is possible at scale, but there's something highly alluring about this idea for the establishment. They believe they can continue business as usual if they promise a technological savior will appear in the future to right our collective wrongs.

The issue with explaining that net zero makes no practical sense is that no one can disprove it. There's no way to prove something is incorrect when it can't be falsified. Net-zero proponents can assert there's no way to prove there *won't* be a seemingly magical innovation in the future that makes large-scale carbon capture possible. One can't effectively argue against a statement based purely on blind faith, but I'll try. The major challenge is that carbon emissions are dispersed throughout the atmosphere and the oceans. Atmospheric CO_2 is the one causing the greenhouse effect, so it's the main focus. In the atmosphere, the CO_2 concentration is over 400 parts per million. Simply put, about 2,500 units of air must be processed to capture one unit of carbon dioxide. So, it takes a lot of processing to capture a little—it's *very* inefficient.

Where did that CO_2 come from? The *additional* carbon in the atmosphere was likely released by the combustion of coal, oil, or gas. For instance, the combustion powered a factory, and the byproduct was CO_2. Ultimately, an enormous amount of energy was used during the release of human-caused CO_2 in the last century. With some intuition and reason, one can imagine that lots of energy would probably be needed to put the genie back into the lamp. Additionally, scaling up and building the necessary infrastructure would also require a great deal of energy and subsequent environmental damage. And there isn't enough time for such hypothetical large-scale carbon capture anyway.[107]

It has taken many decades of fossil fuel emissions to reach the current concentrations of greenhouse gases. This was achieved using the most versatile and energy-dense fuels available. Due to simple physics, any scheme that uses machines or industry to capture these greenhouse gases is likely to fail. It doesn't matter how much money is thrown at the problem—from an energy standpoint, it still makes no sense. The reasons that people persist in working on net-zero solutions are probably the ideologies of modernity, a lack of whole-picture thinking, and the money and careers that are to be had in R&D. Even if we could build seemingly magical machines capable of absorbing greenhouse gases and storing them underground using renewable energy, the scale needed for this to work would be enormous. It would require massive energy and collective effort, and wouldn't address the root cause of continuing to release ever-increasing amounts of greenhouse gases.

The genuinely effective "technologies" that help reduce greenhouse gases are renewable-powered, self-replicating, self-healing, resilient, scalable, and environmentally friendly: We know them as phytoplankton, algae, soil, and plants.

Let's consider geopolitics: Where will the supply for the vast new demand in energy for carbon capture come from? Does this make sense in an age of inflation, energy constraints, global conflict, polarization, and deglobalization? The answer is no.

From energy, financial, and feasibility standpoints, net zero makes no sense. Since I can't use reason to falsify a faith-based idea, I ask you to take a breath and consider what your heart and intuition tell you. Does it make sense for humanity to count on technologies or techniques that haven't been invented to absorb past emissions while current emissions are still rising? Suppose you have a family member who is becoming increasingly sick after using harmful, addictive drugs. Does it make sense to allow greater and greater use of those drugs, counting on a future *genuinely miraculous* cure for drug addiction? Or would you rather see an immediate, gradual phase-out despite the challenging repercussions of withdrawal?

Exploiting fear for profit is easy. There will always be snake-oil salesmen. There's lots of money to be made, regardless of the feasibility of large-scale carbon capture. The net-zero idea won't die easily. It's such an easy way out, after all. Don't change anything about the status quo; build thousands of industrial plants to solve the problem. Easy, right?

Words are cheap, and the words of politicians are some of the cheapest around. If the ruling class sincerely believed in the spirit of net zero, wouldn't they taper off new exploration for fossil fuels? Wouldn't they begin to phase out fossil fuel subsidies? Why do their actions not match their words?

Net zero is a big lie. It's a distraction, an empty promise to keep concerned people at bay. It's such a vast, outrageous lie that few would entertain the idea that someone would distort the truth so profoundly when so much is at stake. The big lie is an effective propaganda technique that carries a certain force of credibility.[108] Every time anyone naively repeats the phrase "net zero," the lie grows stronger.

One of the possible reasons for the propagation of this delusion is the establishment's inability to face reality.[109] Instead of acknowledging that we have exceeded the limits once considered safe, this is reframed as a temporary excursion that will be reversed by "carbon removal technologies" that magically surpass scalability, geopolitical, energy, and resource limitations.

Geoengineering

Why don't we intentionally geoengineer the Earth's climate, since we've been unintentionally tinkering with it for decades anyway? There are traps inherent in following this thought process. One trap is that geoengineering is not a solution. It's a Band-Aid, and like any Band-Aid, it would eventually have to come off. But what's wrong with buying some time when things are dire? The main emphasis of geoengineering mainly focuses on releasing particles into the atmosphere to block sunlight from reaching and heating the Earth. However, that comes with too many potential risks and negative consequences for something that is just a temporary measure and does not address the actual causes.

Ironically, a powerful argument against geoengineering is that if it were successful, it would mask the warming caused by human activities, while those activities would continue undisrupted— business as usual, of course. There wouldn't be direct feedback from the planet as a result of the harm caused by these activities. So humanity would keep digging itself deeper and deeper into a hole—increasing global warming—and would eventually need to expand its geoengineering efforts. It could easily turn into a vicious self-reinforcing nightmare.

Let's consider the practical aspects. A global geoengineering effort would require worldwide coordination and relative peace. The present and future uncertain geopolitical context is not very reassuring. Even something much less consequential, like the COVID-19 pandemic, showed us how troubled international cooperation and coordination are in practice.

A huge challenge in implementing geoengineering is that it requires accurate predictive computer models to minimize weather, environment, and climate disruptions. Today's models are highly advanced, but they still can't accurately predict what the effects could be. There are too many variables and too much uncertainty. Humility should prevail over hubris. That's the lesson of the precautionary principle.

Continuing with the theme of feasibility, geoengineering would require continued funding and resources; the programs would have to continue undisrupted by potential war, terrorism, pandemics, natural disasters, famine, revolutions, political shifts, collapse, and so on. A sudden pause or termination of the program would have catastrophic consequences. It would unleash a drastic increase in temperature, causing worldwide disruptions.

Indigenous Peoples grounded in a holistic understanding of our relationship with the Earth, such as the Saami, stand firmly against geoengineering and insist on tackling root causes instead of seeking "quick fixes."[110]

The wise choice here is to let the living world rebalance itself; actively intervening is only the *seemingly* astute choice. Both are difficult options, but the second is undoubtedly a mistake.

The Living World

Slowly, scientists are rediscovering what has been known by some Indigenous cultures since time immemorial: The biosphere is a metaorganism, just like humans and most complex organisms.[111] Metaorganisms are organisms that host ecosystems of other organisms upon which they depend for survival. For instance, in humans, about half of the cells in our bodies are microbes.[112] It's not entirely clear what specific relationships we have with these microbes and their effects, but it's clear that we co-evolved with them and are in a mutually beneficial relationship.

This insight about humans as metaorganisms may make it easier to visualize how the same is true for the living world. The planet hosts many webs of life. These webs interplay with each other in a way that encourages resilience and the flourishing of life as a whole. The living world is the life force itself. Some scientists refer to this as the Gaia theory, in which they view the relationships within the Earth similarly to those within a human body—between organs, substances, cells, microbes, and so on.

How does this relate to the crises of the living world? Let's consider a medical perspective. Most modern medical science is overly reductive, focusing on isolating variables rather than holistic interventions. Despite its limitations, however, there is value in using a medical framework to assess the state of the living world. If the living world is a metaorganism, there could be ways to evaluate its health. We could identify vital signs representative of its general well-being. The planetary boundaries discussed earlier in this chapter are potential candidates for vital signs. If most of these vital signs are not optimal, we could conclude that the living world is under stress and its capacity for self-regulation and maintaining a stable condition is diminishing. Things could get weird, to put it mildly. Knowing that there could be a point of no return at which things start to get increasingly dire, how much do we want to push the boundaries of our living host?

The insight of Mother Earth as a metaorganism is also encouraging. Like a human body, it has the capacity to rage against the dying of the light.

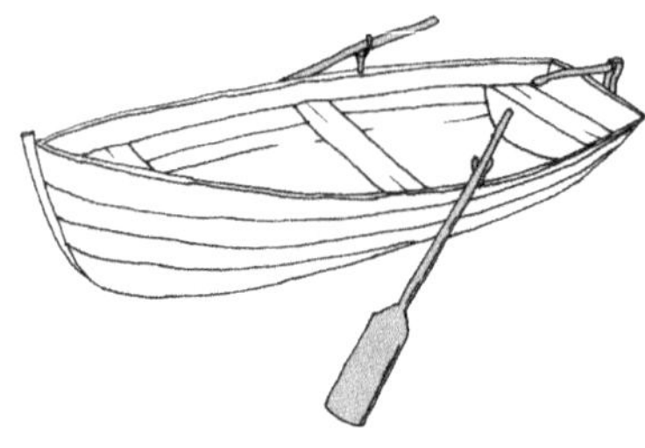

9. Doses of Hopium

Being unwilling to thoroughly consider opposing views is like covering our ears while conversing with someone who disagrees with us. To circumvent this, I've attempted to gather and understand multiple arguments against collapse. Reading between the lines, many arguments seem to come from an impulse that says, "I hope it is not so, because I couldn't stand it if it were true." Many of these arguments may be arising from denial, anger, or bargaining mindsets.

The counterpoints to collapse are not strong, and they lack a comprehensive view. They tend to be based on cherry-picked evidence or managing people's perceptions and behavior. They don't address the arguments head-on, but often focus on tackling mistaken representations—straw men. It is telling and sobering. I keep uncovering mental barriers in those who argue against the unraveling. Most arguments appear to be knee-jerk mental defenses and dismissals, efforts to protect a worldview and its codependent ego rather than providing strong pushback. Perhaps the counterarguments are a form of bargaining with the death of a person's worldview, to keep some part of their identity alive. I will address many of these counterarguments below. This chapter is for those in denial or who are dismissive of this unraveling. Skip it if you're no longer tempted by *hopium*.

Technological Optimism vs. Realism

Hopium: "Technology and AI will save us. Technology creates some problems, but eventually solves those too. The issues that technology creates are best solved with new technology; the solution lies in the exponential improvement of technology. For instance, global warming can be solved with renewable energy. Large amounts of resources are increasingly poured into research, technology, and the energy transition. The costs of renewable energy technologies are plummeting; we'll soon reach a tipping point, and they will become widely adopted. Pessimists underestimate the potential speed of progress in renewable technology. Scientists unanimously believe that civilization can be powered 100% by renewable energy. If nothing else works, geoengineering will save the day. Or nuclear energy will save us—fusion may be around the corner.

Response: The synergistic and complex predicaments we face require a holistic approach. Technology will certainly play a part moving forward, but it's *not* the solution. The belief that civilization's problems are mostly due to greenhouse gases is a simplistic and narrow view. The core issues are the deterioration of the living Earth and the inextricable negative consequences of civilization as a whole. Global warming is just one dimension.

Besides, "clean" energy is not the solution to global warming. The solution is a total restructuring of civilization, society, and our economies. There is an assumption that technology will always outrun the problems it creates. It's a cat-and-mouse game, and betting on the mouse always outrunning the cat is hubristic and naive. Wisdom, community, and moderation are more realistic paths to follow. For instance, quickly scaling up nuclear energy globally would present many challenges due to bottlenecks such as site selection, reactor construction and decommissioning, safety concerns, and nuclear waste management. Our attachments have nothing to do with the dynamics of reality—just as gravity doesn't care what we like, think, or hope for.

"There's no limit to human ingenuity. Humanity always solves problems like these; we've solved them in the past and will solve them again. We can't assume a problem is unsolvable. The alarmists have always been wrong."

There are limits to our ingenuity. The previous chapters describe a perfect storm of multiple crises synergistically reinforcing each other. Our ingenuity is limited by resources, capacity, and coordination. Every technological solution leads to unintended consequences.

Humanity won't likely be defeated, but we'll sustain heavy losses, and we should aim to triage and minimize them as much as possible. Some challenges were solved in the past, but only because people became concerned, discussed them, and took action.

"We are too intelligent to let a disaster like this happen. We are not mold in a petri dish destined to crash after consuming all available resources. We wouldn't overextend ourselves."

We are clever enough to come up with nuclear mutual assured destruction, create viruses and other bioweapons, heat our planet, and drive species extinct. We're smart enough to put ourselves between a rock and a hard place. The citizens of ancient civilizations were also very smart, but that didn't prevent their civilizations from falling apart.

We have already overextended ourselves, and we're experiencing the consequences, as evident in deforestation, pollution, global warming, and diminishing wildlife populations. Humans aren't exempt from the carrying capacity of our habitats. We are more self-aware than mold, but nevertheless the Earth is littered with the ruins of civilizations that overextended themselves. As humans we self-organize through dialogue and other means, which is why the exchange of these ideas is critical.

Misconceptions

"Fossil fuels have an unmatched energy density, versatility, and practicality. And we can't feed ourselves without them. People promoting the phasing out of fossil fuels don't realize that many things come from them, from clothing and plastics to industrial lubricants and asphalt."

Fossil fuels are unmatched, but the fact that phasing out fossil fuels would be extremely challenging and inconvenient doesn't directly address the multiple issues arising from their use. The problem is that we're damned if we do and damned if we don't. That's why this is a complex predicament.

The use of oil, coal, and gas is changing the atmosphere and oceans and heating the planet. It's also powering the destruction of forests and other habitats. At the same time, it's becoming increasingly difficult to obtain enough oil to feed the voracious appetite of modernity. Oil is the blood of modern civilization—not to mention that industrial agriculture is utterly dependent on fossil fuels, meaning that without fossil fuels, there wouldn't be enough food. A complete restructuring of food production (and society) would be necessary to *attempt* to counteract the lack of fossil fuels.

"We will never run out of resources because their cost would increase as their supply decreases, and we would find cheaper alternatives."

We will never extract the last drop of oil or fell the last tree on Earth, because it would become increasingly challenging to access those resources as they approach depletion; that's why their cost would increase. But that doesn't mean the resource won't be functionally depleted. With fossil fuels, this depletion will occur because there's a thermodynamic barrier related to the energy return on the energy invested in their extraction. We will seek alternatives, but they may not be as suitable, abundant, or affordable, and this doesn't address the root causes of the polycrisis.

"Our economies will decouple from the material world and expand in the digital realm without harming the planet."

The server farms hosting these virtual fantasies require enormous amounts of electricity and resources. The physical economy can't be decoupled from the natural world. Perhaps a small sliver of the economy could potentially "decouple," but there will always be a physical footprint. Sufficient effort to decouple a small fraction of the economy would require most of us to face the reality that the Earth has limited resources and to understand the risks of exploiting its resources at an accelerating pace. The "partly decoupling a small portion of the economy" scenario would rely on global peace, cooperation, and no significant disruptions to civilization. But it also ignores that we are already in overshoot.

If we accept that the Earth has limited resources, then humanity would need to rescale the physical economy to respect those limits. Though some observers believe that the decoupling of the economy from carbon emissions is already taking place in some countries, the data they use to support these claims relies on accounting tricks. For example, it often excludes emissions from imported goods, international shipping and aviation, and land-use change when evidence for national-level decoupling is shown.[113] Without such accounting tricks, there is no observable decoupling.[114] It's a big lie, like net zero. Decoupling from the *material* footprint of goods and services is impossible; this can be seen at a glance when comparing global GDP with energy use—both are completely in sync.[115]

"We need economic growth. When people are wealthier, they tend to think long-term and care more about the environment. We can't solve global problems if we're poor."

Affluent people who care about the environment are just a small part of the broader picture. Rich people are part of a totality involving the poor, who extract, harvest, produce, and transport the resources linked directly and indirectly to rich people's consumption. At first glance, it may seem that wealthy communities mostly clean up their environmental messes, but this ignores how a lot of the mess related to these affluent communities accumulates in the places where their money comes from (via neocolonialism), where the goods they buy are produced, and where the materials for those goods come from.

The people who protect the living world most effectively are Indigenous Peoples who live on the land. That's because of their intimate connection and interdependence with those lands. They drink the local water, eat the local fish, and harvest the local plants. Harming or polluting their sources of sustenance makes little sense to them. These people rarely have a lot of money. But some put themselves on the front lines to protect their land from exploitation and pollution, and many have lost their lives in that process.

We can't solve these global problems. There are no straightforward solutions—only responses. Rich or poor, we have tough conversations and choices ahead. We will need resources to adapt to these changes, but currently, many resources are being used for things we don't need.

"We'll be fine. We only spend a tiny percentage of GDP on things like food, so we can double or triple that percentage without issues."

Food production is not merely about money. It is impacted by geopolitics, supply chain disruptions, droughts, floods, storms, heat waves, frosts, wars, pests, energy availability, and fertilizer shortages, among other factors. Countries more vulnerable to agricultural disruptions may destabilize quickly after poor harvests, and conflict could spread easily beyond these regions. And widespread famine can snowball into more extensive and complex issues.

"Humanity will keep carbon emissions under control, so we will avoid the worst effects of global warming. Total emissions have been falling in many countries for at least fifteen years without their even trying. Many countries will be carbon-neutral by 2050. Fusion is around the corner, and hydrogen has great potential. A lot of money is going into these efforts."

The amount of carbon dioxide in the atmosphere is nowhere near under control, and the same goes for methane. Concentrations of both are rising at an accelerating pace. The data showing a decline in emissions for some countries is misleading. It fails to account for international shipping and aviation emissions, and, more importantly, it often overlooks international trade. When country A produces goods for country B, the emissions count against country A, not country B. This flawed accounting misleads us into believing in "falling emissions." In our interconnected world, looking at emissions on a global scale is more accurate.

Without self-restraint, it doesn't matter if we power societies on unicorn farts; if we make products from physical materials and civilization continues to expand into the natural world, the negative impacts will continue. Money can't bend the laws of thermodynamics or create goods out of thin air; everything comes from somewhere.

Energy, Progress, and Resilience

"We have a remarkable ability to adapt and endure; the future will be no different. We've successfully faced past threats to our global civilization. Today's challenges are similar: Past fears of overpopulation and starvation resonate with worries that we will be unable to fuel our civilization without fossil fuels. Past fears over the ozone layer resonate with concerns over climate change. Worries about energy availability are overblown. We've been warned of social and ecological collapse, of peak oil, of nuclear Armageddon, of AI taking over, of cities sunk under the sea. Countless predictions have failed to materialize."

The threat of overpopulation and mass starvation was never actually surmounted. The Green Revolution served only as a temporary measure, as Norman Borlaug, one of its architects, argued during his Nobel lecture: "The green revolution has won a temporary success in man's war against hunger and deprivation; it has given man a breathing space. . . . But the frightening power of human reproduction must also be curbed; otherwise the success of the green revolution will be ephemeral only."[116]

Past impacts to the ozone layer were straightforward to mitigate compared to the enormous complexity of global warming—it was simply a matter of substituting one refrigerant for another. Comparing the two crises is deceiving.

Fracking, offshore drilling, and tar sands are examples of oil becoming increasingly difficult to extract. Peak oil *timing* predictions didn't factor in new extraction techniques. However, the fundamental problems of peak oil and diminishing returns on energy remain.

Many of the threats that inspired the predictions mentioned above never went away. This book explores the current overarching trends and looks into probable future scenarios based on history and the interplay and life cycle of complex systems. As we observe the events around us, we may see more and more dynamics that match the scenario described here.

"Ancient civilizations that eventually collapsed were confined to small regions and had limited and primitive technology. They were fragile, but modernity is different."

Modern civilization is indeed very stable compared to earlier civilizations. Its global range and diverse technologies allow it to withstand much stronger shocks. Civilization is "too big to fail," but no one else can bail it out. The problem with being too rigid and not undergoing healthy pullbacks is that when a system doesn't allow minor shocks to disrupt it, only big shocks can cause real disruption. Eventually, there's a big disruptor and a substantial drop. This is a natural, universal pattern. For instance, a forested area that undergoes frequent small fires is more resistant to fire because burnt-out patches act as deterrents, but an overgrown area that hasn't had a fire in a long time will encourage the spread of fire once one starts.

"Technological and social progress are making the world a better place. Things are improving by every measure. People's lives have improved dramatically. We live longer and are wealthier. Our ability to handle challenges is improving. It's dangerous to cease believing in what raised humanity out of misery and suffering: the idea of progress. Progress has reduced infant mortality, starvation, diseases, and violence. Continued progress addresses the problems of want and mortality. We've done so much to improve the future; giving back any of those gains or stopping progress would be a crime."[117]

The human sphere is embedded in the living planet, but human societies and the Earth are not the same thing. There has been progress in some areas and regression in others, especially in the living world. The dark side is often omitted from rosier pictures, as there are also examples of how human life is declining. The rate of suicide among young people is rising in the USA, Latin America, and Australasia.[118] Mental health issues and autoimmune diseases are also on the rise.

Industrialization may have reduced starvation, diseases, and violence as a *percentage* of the population, but industrialization has put us in our current predicament. Since industrialization increased our population, *more people are suffering from those issues now* than in the pre-industrial past. For instance, in 2023 the number of people under-nourished globally was larger than the total world population in the year 1700.[119] The overall reduction in diseases and violence is debatable, since our larger urban population makes us susceptible to pandemics and chronic diseases (a.k.a. diseases of civilization), and plenty of violence was involved in the world wars (and a sequel is always possible). War-related deaths from the post-industrial era eclipse the prior war-related deaths.[120] In terms of diseases, we can home in on pandemics as an example: If we compare pandemic death tolls before industrialization (e.g., Black Death, Columbian exchange) with those post industrialization (e.g., third plague, Spanish flu, AIDS, COVID-19), the total number of deaths post-industrialization is also higher.[121]

The so-called problems of want and mortality are not solved through external conditions. Ancestral traditions offer guidance on navigating these challenges—for instance, through internal reflection, rituals, gratitude, understanding, and conscious presence.

Ultimately, industrialization has subsisted entirely on a one-time bonanza of fossil fuels, and alternative energy sources are insufficient. Even if there were a hypothetical unlimited energy source readily available, the degradation of the foundations of the living world would continue unrestrained.

There is hidden fear lurking in those who claim the past was horrendous. When fear controls us, we unintentionally become its servants and propagate fear onto others. Many wonderful things about the past get obscured by fear or cultural narratives.

"We still have time to reform or transform civilization."

This is unlikely. How much time is left? Is humanity really attempting to change? Wide-angle observations don't suggest this. And if we're not fully mobilized already, what makes us think we'll do it in time? Are we procrastinating until the last minute? Such a complex challenge has to be met head-on with substantial effort. Civilization can't turn on a dime. Are we just shifting the goalposts? Tackling the predicament head-on to aim for a softer fall would entail a complete restructuring of our societies. It would involve a drastic reduction of resource use, which would require voluntary simplicity and a complete overhaul of everyone's lifestyles, among many other things.

"We're making actual progress: The public is more engaged, the movements have grown, and we've had significant victories."

The raw data isn't lying: The living world's vital signs haven't improved significantly (e.g., CO_2, ocean acidification, land-use change, biodiversity, freshwater, nitrogen, and phosphorus). The overall trends remain unchanged, and the material needs of civilization continue to grow year after year.

Mindset, Morale, and Emotional Responses

"Life before the industrial age was terrible—worse than any future we could fear. Without modern progress, life would be horrific."

That's a purely emotional argument, and one arising from fear. It distracts us from the fact that industrial modernity, as we know it, is temporary. If life before the industrial age was so terrible, why didn't those who lived through it repeatedly emphasize how horrible it was in their stories and writings? How could they create beautiful art under such circumstances? Why would they endure such challenging conditions if their lives were so miserable? What for? Why didn't they just succumb to suicide and diseases?

Assuming that humans lived terrible lives before industrialization is a narrow-minded view. It's the reverse projection of idealizing the modern age. Thinking that the past was so bad is likely a projection arising from a failure to see the dark side of our present—and of our possible futures. The mind may mask or justify the shadows of our time so we can feel good about ourselves. It may deflect criticisms and distract us with a decoy: "I did nothing wrong. Look at them over there; they did much worse things."

"Without widespread optimism, progress will be slow. If we become convinced the world is headed for doom, we might as well give up. Negativity and pessimism will paralyze, demoralize, and discourage people from working on solutions. Hope is needed to keep people engaged and motivated.[122] The ideas that we lack agency and that it's too late are now a primary obstacle.[123] People may be too quick to gravitate towards the certainty of defeat.[124] We need action, but these attitudes turn people away, making them think they can't do anything about it."

Arguments like these are about managing public morale, as if we were in a war. Metaphorically speaking, journalists involved in propaganda to win "hearts and minds" shouldn't argue with military officers about the state of the war effort, because journalists can't discuss the military's ability to win or lose without a substantial bias.

These public-morale managers would surely have objected to Winston Churchill's exhortation, "I have nothing to offer but blood, toil, tears, and sweat." That is not exactly a promise of eternal bliss and sunshine, but it's real.

Our attitude influences our capacity to do things, but it has limits. We may wholeheartedly believe that we can stop the Earth from spinning, but that won't make it happen. The systems around us have enormous inertia; things will change, but there's no escape from some reckoning.

There's the issue of public morale—hearts and minds—and the issue of what is happening. We can't afford to confuse the two. A severe diagnosis may discourage some people, but a doctor doesn't base a diagnosis on what motivates or cheers up a patient. Good doctors first seek an accurate diagnosis. Only then can they go about informing the patient skillfully.

Within the paradigm of modernity, taking action may seem pointless if a breakdown is already built in. Since we are mortal, we are bound to die—but that doesn't mean we will spend our lives lying in bed, waiting for death. Wise people plant trees even if they will never benefit from their shade. Accepting the reality of an unraveling doesn't mean giving up. People respond to challenges against all odds, as we're wired for survival. In many ways, denying the severity of a diagnosis can be a massive obstacle to healing.

If we remain within the paradigm of modernity we will be sitting ducks, limited to ineffective responses and false solutions. Agency and power are found outside the box. More than attractive visions of the future, we first need people who can engage with the reality of the situation. Refusing to see the complexity and severity of our predicament is a recipe for tragedy. We can't afford to lie to ourselves.

This book is not an argument for defeatism but a call for action regardless of the odds. It's also a plea for recognizing that some current efforts are not working and are likely to fail. We may be desperate for actionable guidance, but before we act, we need to clearly and accurately comprehend what is happening, no matter how scary or terrible it may seem.

Feeling paralyzed, discouraged, and demoralized may be a necessary phase that some people must go through as a catalyst for powerful action—action emerging from a deeper place than hope.

Sometimes, one must let go to fall upon a more solid foundation. Powerful action, spurred by love, wisdom, and deep insights, doesn't require hope. History has plenty of examples of this.

We shouldn't prevent ourselves and others from experiencing certain emotions. Powerful action doesn't come from avoiding difficult emotions; it comes from connecting deeply to those emotions and transforming them in the service of life. There's a place beyond despair where powerful new beginnings may emerge, but by trying to avoid despair, people fail to reach this place.

The role of engineers, scientists, researchers, and other people who highlight these issues is not to be cheerleaders. Attempts to silence and suppress their views are merely a way of managing expectations and propping up narratives.

It's time to move beyond the talk of agency, hopes, and fears. It's time to talk about what truly matters to us, the world we'll leave to our descendants, our roles right now, and our power as co-creators of our reality. There are enormous forces in motion all around us. Recognizing these realities doesn't need to paralyze us any more than it would paralyze us to acknowledge that the sun has a finite lifespan.

"Fear and panic will paralyze people and become destructive."

Suppressing the severity of a situation involving all of humanity and the living world is wrong and counterproductive. What gives anyone the right to withhold crucial information? Suppressing information causes mistrust, confusion, and division, making effective coordination much harder. It also prevents people and communities from making informed decisions and acting in time to improve their circumstances.

Fear may initially cause a freeze, fight, flight, or fawn response, but these reactions can give way to wiser responses. That being said, panic would be counterproductive. That's why mature people should exemplify ways of responding—responses based on a deep foundation, rather than on fear and selfishness.

"These ideas promote letting go of the steering wheel."

If we look carefully, we may find that collectively, from a ground-level view, we have limited use of the steering wheel to begin with. For instance, in recent years, no one could have predicted the impacts of the internet on society. There have been and are active attempts to influence its development; however, the internet continues to shape society in emergent and complex ways.

If humanity has a good grip on the steering wheel and can make meaningful changes, why hasn't it made them already? Even the presidents and CEOs of the most powerful countries and companies experience huge limitations on what influence they can have globally. This is not a call for letting go of our agency; it's about pointing out the many trends around us that possess massive inertia and a life of their own.

"I reject the defeatist attitude. Humanity shouldn't just say "nothing can be done," lie down, and give up. Defeatism may be just as dangerous as denial."[125]

This book does not advocate giving up, but rather encourages the realization that the current path is self-destructive. We will suffer severe losses, but humanity will undoubtedly persist. There is an essential difference between pessimism and realism. Refusing to see the reality of the difficulties we face is dangerous. Blind optimism makes us complacent. Of course, giving up makes no sense. But there are things we should definitely give up, such as certain paradigms, beliefs, habits, projects, efforts, and lifestyles. Either we give them up, or the era of consequences will forcefully take them away in a manner that causes more damage.

"The problem is a scarcity mindset (as opposed to an abundance mindset). A positive attitude will overcome the issues, but pessimism won't. If we see existential problems everywhere, we won't be able to prioritize our efforts. These ideas stem from an extreme fixation on the negative news. It's also a failure of imagination; we're losing because rich and powerful people can't or won't imagine a better way. This is your life, and it's ending one minute at a time. Do you want to spend it afraid of doomsday, or make the best of it?"

Attitude is essential. Under challenging circumstances, it's best to live with a paradox: Confront the brutal reality, and keep faith that we will prevail or that what we do matters. That's not Hallmark-card wisdom. That's the paradox that enabled James Stockdale to survive horrific conditions as a prisoner of war in Vietnam. When asked who were the first among the POWs to perish, he responded: The optimists; they died of a broken heart. The pessimists followed shortly after.[126] Balancing the brutal facts while maintaining an unwavering conviction is essential for resilience. Courage is not the absence of fear; it is a heart-based impulse despite the fear. This talk of positivity, abundance, and growth doesn't mean much to people who have endured ordeals against all odds.

A failure of imagination is not what's preventing us from "solving" our metacrisis. That's a very simplistic and naive statement. This predicament involves psychological, cultural, geopolitical, technological, operational, economic, and political aspects, among other challenges.

How each of us lives our lives is very personal and unique. We're free to do as we please to an extent. That being said, we live at a unique moment in history when our actions, however small they may seem, could have significant impacts.

"God wouldn't let this happen."

Many religious myths warn of catastrophes when humanity is out of balance—and many of those myths involve cataclysms of epic proportions. No matter what our beliefs are regarding God, dismissing these ancestral warnings is hubris. If we believe God would intercede, then those actively responding to these crises could be expressing the direct and indirect manifestations of God's will. It's possible that God's intentions are being expressed through these people in their own ways.

"The responses suggested for mitigation and adaptation, such as phasing out fossil fuels or significantly reducing consumption, are problematic and inconvenient, so the issues are probably a hoax."

This is a widespread sentiment among people who deny our civilization's predicaments. It's a mental defense against the emotional implications of the issues, and sometimes it's even a result of propaganda. This denial can best be resolved on an emotional level through skillful, gradual, and gentle one-on-one conversations involving active listening.

"I don't see the crises. These minor challenges are so overblown. Humans will adapt. This is hysteria. These arguments are a projection of spite, anger, or depression. These people are just looking for obstacles; if one is removed, they find others."

Many of the crises we face aren't immediately obvious. They require weaving together trends, while imagining how they interact and reinforce one another. Yet anyone can see the signs if they understand the themes. The signs may not be widespread if we live in a relatively stable nation rich in resources, but we may still see indications of economic, ecological, and social deterioration. Our socioeconomic status can act like a cocoon that insulates us from the surrounding impacts—the crises aren't hitting everyone and everything simultaneously or with the same intensity.

The warning signs of our predicament are often pointed out by engineers, scientists, academics, researchers, and others who have studied these challenges for years or decades. These people are generally not the emotional type, and they also try to avoid rocking the boat. If a substantial number of them are sounding alarms, they must believe it's serious, as they are putting themselves in a vulnerable position—perhaps risking becoming an outcast, losing a job, losing friends, or harming their reputation. Many people raising these issues are frustrated by the greater society's reliance on distraction, denial, and token efforts that are ineffective or counterproductive. We face significant losses, and grief, sadness, depression, and anger are all emotions that may visit us as we realize this.

Political and Social Challenges

"Warnings are only helpful if something can be done to prevent the outcomes warned about. If a breakdown is baked in, why should politicians and industry support new technologies and ways of being?[127] *If it's all hopeless and there's no point, why not continue razing the forests?"*

The upcoming suffering can be minimized or amplified depending on our actions. This warning is about responding and preparing for the age of consequences. The living world is not ending. However, the way our societies function will become "simplified" and vastly different.

Politicians primarily care about holding on to power. We can't expect lucid responses from them as politicians; any genuinely helpful response would come from their hearts, not their desires for status and power. We shouldn't hold our collective breath either way.

Industries can profit by providing appropriate technologies and goods, but ultimately, industries have no heart. Decision-makers in these industries should play a role, but their motivations would need to come from something more transcendent than money and power. Slaves to money, power, and control can't possibly be helpful in these times. Continuing down the path of overexploitation will make things harder for all living beings on our planet and for everyone's descendants.

"You're buying into a narrative that profits some people. Who benefits from these existential fears?"

There's a lot of money to be made from government contracts and people's fears. Opportunists will profit from this. And while some people may think that's wrong, it doesn't mean there is no predicament.

"It's all just an excuse to implement a one-world tyrannical government and take away our rights."

Global, centralized enforcement of "solutions" sounds like tyranny, and it is. There already is a trend towards totalitarianism in many countries, as evidenced by pervasive surveillance, the use of "terrorism laws" to suppress dissent, and the criminalization of protests and strikes. An attitude of "Daddy state knows best" will worsen the situation. The same desire for control at the core of totalitarianism is also significantly responsible for our predicament. Often, people seeking control use crises to consolidate power, but that doesn't mean these crises are manufactured. The polycrisis is real.

"We have the solutions right now; we're only missing the will. The obstacles are political. The public is not the problem; the problem is powerful vested interests."[128]

It's partly true that "we have the solutions right now." In theory, we can solve war, extreme inequality, discrimination, and so on, but things are not that simple in practice. Anyone reducing this metacrisis to a single opposition, aspect, or area is viewing things through a very narrow understanding. When people propose "simple" but false solutions, it hurts genuine efforts.

Miscellaneous Critiques

"We can't predict the future with certainty."

Unsustainable systems end; societies divorced from ecology lack longevity; and our species can't survive alone. Arguing against that requires extraordinary evidence.

Predictions will always come with high uncertainty, especially those projected further into the future. But the lack of absolute certainty doesn't mean informed predictions are not helpful. As they say, history doesn't repeat itself, but it rhymes. If we understand these rhythmic patterns, we may foresee scenarios that are likely to happen. As human beings, pattern recognition and predictive skills are undoubtedly our stronger abilities, so it makes sense to use them for *some* guidance.

"The planet doesn't need to be saved."

The living world will prevail; it's more resilient than humanity and civilization. It has endured cataclysmic asteroid impacts that make nuclear war look like child's play. Modernity and humanity, however, are not as resilient as the living world and will suffer great losses. Nevertheless, human extinction in the near centuries is very unlikely.

"This is an argument for degrowth. Without economic growth, everything will be worse. Degrowth is political suicide, and it will never achieve popular support."

This book is not an argument for degrowth as a political strategy. Having said that, due to internal conditions and external triggers, a decline *will* occur eventually, regardless of our actions. Given this, there are things we can do to soften the fall and choose a place to land. Keeping the public out of the conversation will reduce support for productive approaches as disruptions grow. There is much more to life than the economy. The foundations on which humanity stands are at risk.

"Humanity can be fed using organic agriculture or permaculture."

That may be possible in theory, but shifting the current agricultural systems would require a complete overhaul. It would require substituting a lot of heavy machinery with human labor, and a major portion of the population returning to agrarian lifestyles. Our societies would have to change radically—such that they would be unrecognizable to the modern eye. It would take time, coordination, global cooperation, and many resources, all of which we lack. Further, this apparent solution overlooks several other challenges and fails to address the core preconditions and triggers of our polycrisis.

"A mass awakening will happen soon."

There won't be a mass awakening that prevents this reckoning. There will be a *late*, partial, rude awakening as the big dominoes fall, but it won't stop the crises from unfolding.

No one's coming to help us during this live experiment. Neither God, nature, nor any other entity will save us from ourselves. We're on our own. This is for us to deal with. That's the way these things go. We are expressions of life, figuring out how to be with ourselves, our shadows, and the world around us. The sooner we realize this, the less suffering we'll create.

"Nature is nasty and brutish. The tree huggers trying to help nature don't realize how vulnerable they are and how ruthless nature can be."

The natural world is neutral. It's not trying to make us suffer. If the idea of a malicious nature resonates with us, it could be a projection of our insecurities and the unease arising from the many aspects of life that are beyond our control. Or it could be a cultural narrative.

Life tries to preserve itself. It is neutral at a "personal" level, but collectively, it's trying to survive and thrive. Mutual aid has allowed life to flourish as it has. Of course, there is also competition. The occasional flash of indifference from nature highlights how important it is to play nice with it. Not only is modernity self-destructing, but it is also undermining the source that sustains humanity.

"These ideas are anti-human. They seem to imply that the world would be better off without us. They frame humans and our activities as evil."

The planet wouldn't be better off without us. Reality is not simply black and white. Humanity and its actions are not evil, yet humanity can do terrible things. It's important to recognize causes and effects. Avoiding the recognition that humanity has participated in something harmful is misguided. Are we able to accept that we can make—and indeed, have made—huge mistakes?

Socioeconomic Critique and Hypocrisy

"People only worry about this because they are well-off and don't have to worry about their basic needs. They are spreading privilege-induced guilt."

That is an attack on the messenger, not the message. People who aren't meeting their immediate needs may worry less about these broader crises, but personal worries are one thing, and the existence and severity of our collective challenges are another.

"The people ranting about these issues are hypocrites. They use modern tech, plastics, and fossil fuels, eat meat, and act like everyone else. Their actions don't match what they're saying."

Many of the people who are on board with tackling these challenges do not live according to their beliefs. There's a visible disconnect. However, people live within systems. There are taxes, laws, bills, building codes, bylaws, employer expectations, licenses, permits, traditions, debts, social and family responsibilities, et cetera. Most of us are somewhat trapped in these systems; it's difficult and at times impossible to avoid participating in them. This is also a challenging aspect of our predicament—this state of "capture" hinders adaptation. Calling people hypocrites might be compared to calling a child who suffers from child abuse a hypocrite for living with his parents.

Those highlighting the need to tackle our metacrisis are also struggling with the contradictions within our world. If we think the best we can do to respond to our problems is to focus on minimizing our individual environmental impact, we haven't truly grasped the scale of the predicament.

Part 3: Responses

10. Personal Responses

"God grant me the serenity to accept the things I cannot change, courage to change the things I can, and wisdom to know the difference." Reinhold Niebuhr

The first step in responding to our predicament is accepting that we're in one. The next step should be to respond appropriately and prepare for the age of consequences. We should learn to live with this context, let it transform us healthily, and honor our roles as we move forward. The following are suggested approaches for how we could respond to this unraveling.

Inner Responses

OBSERVING OUR PROGRAMMING

Those who wish to respond directly to our challenges should embark on an inner journey. Personal change is needed for genuine outer change. Without self-awareness, we remain trapped by the habits and beliefs of our individual and collective subconscious. Our responses will likely be misguided and inappropriate without an inner alignment to life. This inner journey also helps mitigate ineffective strategies, intolerance, group conflict, and emotional issues.[129]

However, focusing only on inner change before responding is unnecessary. Nothing exists in a vacuum. Inner and outer conditions are intertwined. For some, starting with an *inner* change is natural. Working simultaneously on *inner and outer* conditions will be a better approach for others. And for some people, focusing on the external situation will be the catalyst for inner transformation. We can view inner and outer work as the balancing and pedaling required to ride a bicycle.[130] If we do both simultaneously, they reinforce each other, and we'll maintain forward momentum. This inertia facilitates further inner and outer work. If we focus only on one or the other, we're more likely to disengage, fail, or burn out.

Ultimately, there must be a component of inner transformation, or we'll continue to propagate our imbalances and those of the collective. However, if inner work is our sole focus, we'll refrain from sincere and genuine creation by not engaging with the external world. Our thoughts, speech, and actions reflect our internal world, which is why external transformation entails an inner transformation. External changes are fragile in the face of difficulties and hardships without an inner transformation.

We can't be wise participants in this merry-go-round unless we have a rudimentary understanding of reality, what we are, and our relationship with the world. A good place to start our inner journeys may be to reconnect ourselves with the joy and pain of our own emotions. Another possible starting place is exploring what ideas compel us deeply. There's no substitute for self-knowledge. Each of us should observe ourselves and contemplate the patterns we identify within. Through these observations, we may shed light on how our thoughts shape our emotions and how our emotions influence our thoughts. Understanding ourselves leads to a deeper understanding of others. Discovering our nature-and-nurture programming allows us to see the traits we share with other people.

BEYOND LEARNED HELPLESSNESS

To respond appropriately to this metacrisis, one habit we should let go of is learned helplessness. We sometimes succumb to a state of learned helplessness when we repeatedly find that we have no control over adverse conditions. To move beyond learned helplessness, we must practice improving our conditions in order to realize that some things are actually within our control. Modernity helps domesticate us into learned helplessness because it cuts our direct ties to the land and acts as an intermediary—we meet our desires and needs through civilization. We're expected to participate in a pre-designed roadmap created for us by the system. But when the system itself is on a path of self-destruction, messing with the stability of the biosphere, our playing by its rules is irrational and even insane.

Going beyond learned helplessness means doing what we can, given our circumstances. It means taking responsibility for our share, and outgrowing the mindset that something or someone will fix things for us. It takes maturity to realize that, at some level, you're on your own—and it's up to you to co-create the world. The power lies in getting things done ourselves, not in pleading and bargaining. With great *discernment* and without underestimating ourselves, we must stop giving our energy to the things we can't change and start giving it to the tangible things we can.

OUTGROWING THE VICTIM MENTALITY AND US VS. THEM

The "us versus them" mindset is toxic and will lead us down the wrong path. It is something that should be discarded. Our power lies in letting go of the victim mentality and refraining from painting the rich, the "elites," the "evil cabal," and the "decision-makers" as the perpetrators in this tragicomedy.

Of course, a small fraction of people have had more say than the majority in determining civilization's trajectory. Still, our entanglement means we're all perpetrators and victims within civilization. Almost all of us have participated in this act, and blaming a portion of the actors for the overall play is overly simplistic. It's also disempowering and unwise. Blaming and scapegoating certain people will alienate them and ultimately backfire. True agency lies in seeing and understanding how we play the role of both victim and perpetrator; genuine power is found in how we respond to this awareness.

SEEING THROUGH THE ILLUSION OF CONTROL

The nature of agency and free will is counterintuitive because we must understand what forces limit and shape our agency before we can exercise genuine agency. It's likely that the more agency we believe we have, the more deluded we are about what truly pulls our strings and how the forces influencing our reality operate.

We're not chained to a predetermined future that is impossible to deviate from. However, as individuals we tend to significantly overestimate the degree to which we have a sovereign, separate, and uncaused will. And if we aren't even in control of ourselves, how can we ever control external conditions? If we look closely at ourselves, we'll find that our circumstances have greatly shaped us. Of course, we shape our circumstances as well, but many of the most significant influences on us—like the lottery of our birth—were and are outside our control.

Reflecting on this interplay between our context shaping us and us changing our context may help us imagine how the same happens for everyone else and for the world more generally. Being enmeshed in this interplay means we exist in a spiral of events and responses, events and responses, and so on.

The majority of us are caught in our circumstances. We've lost ourselves in them and haven't paused and realized that we are pawns of our context—this goes for kings, slaves, and everyone in between.

There's a degree of genuine agency in this spiral of events, but it's only available when we *step back*, ground ourselves, remember we're in this spiral, and respond accordingly. The paradox is that actual agency lies in understanding our entanglement and aligning ourselves with natural forces and order. Agency requires practicing discernment and wisdom, and paying close attention to our emotions, thoughts, senses, and the natural world.

RECONNECTING WITH OUR INTUITION

Over-reliance on *genuine* reason has not led to the metacrisis. We've never been entirely rational—in fact, we can't become fully rational and should never strive for that. We are reflections of the universe, and the universe is not rational.

Collectively, we over-rely on self-centered pseudo-reasoning, which is leading to our disintegration. Acting against our own well-being is irrational, but that's what humanity is doing in the broader sense. We are rationalizing the desires and aversions of our narrow-sighted egos.

How can we navigate an extremely complex environment in which our thoughts and actions can lead to unpredictable consequences? No fail-safe navigation aid exists, but relying only on reasoning—given our history of narrow-mindedness—is a recipe for self-destruction. The world is unreasonable, and reasoning our way through it makes no sense, but many of us haven't received this memo.

We have to stop clinging to ways of thinking that reduce and oversimplify reality into problems to be solved, win-lose games, and equations that treat the world like a handful of variables. This narrow, reductionist way of thinking helped create many of the predicaments around us. When we over-rely on our ego and reasoning—which often amounts to *rationalization* rather than actual reasoning—we neglect other, more holistic and broadly attuned ways of perceiving and processing reality.

We should get in touch more deeply with our creativity, imagination, intuition, senses, and the natural world. The world is an extension of our brains and vice versa. Moving beyond the reduction of the world through mental models is key in this age.

UNLEARNING MODERNITY'S PROGRAMMING

One of the most powerful interventions we can pursue is emancipating ourselves as individuals from the paradigms of modernity and civilization. We need to let go of the mindsets of domination, control, superiority, unsportsmanlike competition, separation, and not being part of nature. These drives arise from deep fear, insecurity, and a refusal to accept life on its terms and love it as it is. Part of letting go of these paradigms is withdrawing from collective delusions and expectations. The main delusion is *separation*, the belief that we can harm someone or something without harming a part of ourselves.

We're primates, so we have powerful instincts regarding status. Most of us have been socialized in ways that lead us to desire money, power, or status symbols. The situation we find ourselves in requires us to be aware and try to liberate ourselves from that programming.

Unlearning what modernity has programmed in us is not straightforward. It requires profound observation and inquiry. The primary reference point is the natural world, for it is a sacred expression of the universe's inherent order and forces. The natural world is the ultimate teacher and a direct source of wisdom. Ancestral and Indigenous teachings on living lives closely aligned with the order and forces of life are secondary reference points, and there's a lot to learn from these perspectives.

Shared collective stories shape our worldviews and tell us about who we are, our responsibilities, the kind of agency we have, our options, and the future that awaits us. These stories are compelling. They powerfully shape our perspectives and our lives. We must notice the power that these stories have over us and search for new stories if we find the old ones obsolete as we face the metacrisis. New, better-suited stories can help empower us and allow us to see the meaning in our lives.

The predicament we're dealing with has roots inside of us. We must be honest about how the drive for security, control, competition, domination, and separation act within and through us. We should examine and deal with these drives within ourselves as we address them outside ourselves. We have the most influence in our own house, after all.

LIVING A SIMPLE LIFE

Some of us need to withdraw from society to examine our worldviews; we shouldn't discourage people from retreating to their own spaces. Real innovation and powerful change often arise from quiet time and solitude. Life is simple. Focusing on meeting a few basic needs can provide us with space, time, and energy for healing and transformation. We should start embracing voluntary simplicity, not in a romantic way that frames it as an easy life, but as a genuine recognition of the recalibration that can happen when we retreat from the hustle and bustle of modernity. As we collectively endure a "long descent," another benefit of voluntary simplicity is that the closer we are to the ground, the softer the fall will be.

Navigation Aids

RECALIBRATING OUR COMPASS

Misguided ideas, habits, and mental pathways may naturally disappear through inner work. One way to initiate a virtuous cycle is to use better navigation aids and transform our context to more fully align with the expressions of the universe manifesting through us.

We're individually and collectively entangled in modernity's highly complex unraveling. We can't use the same paradigm that led us to this predicament for navigation. It would backfire.

In this context of entanglement and unpredictability, we should navigate with the assumption that fog clouds our ability to foresee the full consequences of our actions. Acting within this uncertainty can feel unfamiliar and paralyzing. How can one be in service of the good in this context? The answer is unclear. There's no guarantee that one action or the other will ultimately lead to a better world. It's not for us to decide how things play out. The best we can do is align ourselves with the forces and life-aligned insights that have remained relevant throughout time.

None of the following navigational aids aims for specific outcomes. Someone living or acting from these aids no longer needs to seek outcomes, and someone without a need for outcomes will have easier access to the aids. It takes time, effort, and practice to reorient ourselves and recalibrate our compasses. The following pointers are deliberately not fully delineated or clarified. Following these signposts amid the noise and distraction requires great discernment and care on the part of each individual.

LOVE

Love is an obvious signpost and a powerful life force. It can manifest in courageous and aggressive ways, like a mother bear protecting her cub, and it can also be humble and self-sacrificing. It can make us feel immense grief and sadness, like a whale mourning her dead calf. Love is a real and profound recognition of our kinship and interbeing.

A simple way to know if you are aligning with love is to ask yourself sincerely if you are operating from fear or from love. We are acting from a fear-based perspective when we act because we fear specific outcomes. Fear can also manifest as the fear of failure, embarrassment, exclusion, losing privileges, and so on.

If we operate mainly from a place of love, we may sometimes contemplate fear-based ideas, but love won't let us become fully paralyzed or manipulated by them. Fear is an aversion, and love is an attraction. They're not mutually exclusive, so noticing which manifests more frequently in our lives is helpful.

BEAUTY

Beauty is another signpost, but it requires a lot of discernment. Beauty is not as straightforward as other guides because it has objective and subjective components. The beauty being referred to here is the beauty aligned with life. This beauty is raw and real. It's always in flux. It's natural and emergent instead of contrived.

TRUTH

Humans are prone to deception. When our egos are tightly bound to specific ideas, the ego filters contradictions and counter evidence as an instinctual self-preservation mechanism. We don't need others to deceive us. We can easily deceive ourselves.

Lies, ignorance, and deception are highly corrupting. It's wise not to spread them. Failing to tell the truth in our modern context, where lies propagate easily, is quite short-sighted. Deep down, lying to someone else is a form of self-deception, both in the "lying to yourself" way and in terms of "shooting yourself in the foot."

Many people serve Mammon—money or wealth—rather than the truth, so Mammon becomes their master. Truth is a much more transcendent force. It has no agenda or interests; the truth *is*. Aligning ourselves with "what is" is not only the wise thing to do but also a foundational step in finding true agency.

HEALTHY DETACHMENT

Our egos are crucial for our self-preservation, and yet despite being entangled in an unraveling that directly threatens our survival, in this instance our egos are often more of an obstacle than an aid. Our egos not only obscure aspects of our reality, but they often channel our efforts into narrow-minded pathways that lead to worse realities for everyone. But how can we healthily detach from our egos?

Having personal goals and strategies is not wrong or something to avoid entirely. The issue is that our goals may be misled or co-opted by our egos, fears, desires for control, and so on. As we work towards our shared goals, guarding against these corruptions is crucial. We must actively inquire about the underlying and hidden motivations and impulses driving our actions. Doing so requires lots of attentiveness, discernment, wisdom, and inquiry. This is incredibly hard, but focusing on listening and expressing our nature with as much fidelity as possible is a good start. If our goals and strategies are not deeply steeped in nourishing our hearts and minds, they are neither wholesome nor strategic.

The most straightforward way of keeping our egos in check is to let go of our attachment to outcomes. It's so counterintuitive to the modern mind. Why would letting go of outcomes be a good thing? It's because seeking outcomes got us here. The road to hell is paved with good intentions.

We don't need to visualize a destination. The more we try to crystallize our preferred visions of the future, the more we project our fears and shadows onto them. It's a vicious self-fulfilling prophecy. It's hard to explain briefly how this is so. I could write about the coincidence of opposites or how "the ends justify the means" thinking leads inevitably to the abyss. Instead, I ask you to contemplate how our personal resistance and aversion to reality create suffering for ourselves and others.

To co-create a better future, we should accept and embrace the present, the process, and the journey. And to fully embrace them, we will need to let go of our desired outcomes. This is a way to partially detach from the ego. Only then will we know if what we are doing fully resonates with our whole being and if we are being true to our unique, deep motivations.

Embracing the process means acting in a state of acceptance and peace. The energy of joy or acceptance flows into anything we do. But if we can't accept or enjoy the process, there is a conflict within. Actions in that context don't come from a place of integrity, but from a narrow and fragmented perspective. When we are in that state of conflict, we fail to take responsibility for ourselves, which is the most important thing we can do.

We should resist the modern illusions of certainty and stability. We need not cling to the desire to shape the future in a certain way. Our rigid ideas of what the future should be are obstacles to the better futures that could emerge if we were to create some distance from our egos.

It's important to realize that all our actions and inactions are world-building. In our unpredictable and complex context, actions can have disproportionate effects. Our power to shape the future is outside the grasp of our minds. Knowing that, there's no need to become trapped in despair; our actions and inactions influence the world in ways we can't possibly understand. Acting from this awareness is true power.

LIFE AND THE SACRED

Life is a miracle. Many of us have taken it for granted for so long or have suffered so much that we either forget or never realize how extraordinary and miraculous life is. Gradually, for countless generations and in many parts of the world, the shared sense of the sacred has shifted from the terrestrial to the celestial. There are many beliefs about the sacred, but conscious thought is not necessary for witnessing it. Careful, open-minded observation is enough to realize that Earth is our mother and grants us the gift of life at every moment. Being aware of the sacredness of life and of the Earth may require contemplation and inquiry, but no faith is needed—the miracle is self-evident.

To align ourselves with reality, we must ground ourselves in the sacredness of life. With a deep understanding of our relationship with life, our responses to the challenges of the polycrisis will emerge naturally.

Taking life and the Earth for granted, without gratitude, is a big mistake. Humanity is not the foundation of life; life is the foundation of humanity. There's a hierarchy, and if we don't recognize it, we won't last long and we'll cause great damage as we tumble.

Many modern societies are out of touch with the sacred; all the more reason we should rekindle our sense of the sacred world. Spiritual and religious teachings may provide rough initial guidance, but there's no truer guidance than that of the source. Nature is the sacred miracle. We need to honor it. Falling back in love with nature is crucial in these times. There's no firmer grounding.

Most societies have transitioned from walking on eggshells to please the gods, to a relaxed dance with creation, to thinking we're the center of the universe. These days, it's all about what *we* want: If we can do it, it's our God-given right. But it's time for us to align with the sacred.

We should create our own personal space within which to connect positively to a larger reality. Practicing prayer and rituals can be helpful. We should cultivate a sense of belonging, connectedness, and a responsibility to a larger reality that cuts across generations, groups, and living beings. We should contemplate what we have in common with our ancestors, people living in the present, and future generations. We should feel good about being intertwined with the Earth and the universe.

As we handle massive disruptions in all areas of life, it will become increasingly necessary for us to reconnect with the notions of spirituality and the sacred. Spirituality puts into context our relationship with nature—with what is not human-made. It suggests models for self-realization or going beyond ourselves as individuals.[131] As we face the age of consequences, spirituality will be an essential source of resilience, meaning, and inspiration.

INTERBEING AND CONSCIOUS CREATION

In large part, the dominating and self-destructing paradigm of separation has brought us to this point of crisis. The way forward now is through the insight of interbeing, which allows us to get closer to what is good.

In a self-terminating paradigm, the ends justify the means, but in the paradigm of interbeing, the means are an end in themselves. To practice the insight of interbeing, we have to be compassionate and humble. This requires understanding and contemplation, as we recognize our limitations and consider that we may be wrong. Through practicing gratitude, humility, and reciprocity, we can embody the insight of interbeing.

With the insight of interbeing, we recognize that everything is linked to everything else. Conscious creation is simply the awareness that every thought, speech, action, and inaction has its own karma and transmits different energies into the world. We co-create our reality at all times, and when we are conscious of what we spread, we become more mindful of our part in the creation process.

The actions aligned with the insight of interbeing reinforce the notion that we are all connected. They *integrate*, they don't separate, and they encourage the emergence of wisdom in all of us. They are for all of humanity, not just *my* country; in fact, they are not just for humanity, but for all living beings.

Inner Work

BEING DIFFERENTLY

Our modern crises did not arise from a lack of information. In many ways, they arose from our way of being. Collectively, we reinforce in each other the notion of separation in many ways: how we see the world, talk, relate to each other, act, and so on.

We must practice different ways of being if we want to better align with reality. *Being differently* is a necessary step for showing up differently. It's not enough to understand our predicaments intellectually. We need to develop the capacity to connect and relate differently to the daily realities of life, the larger reality, and our grief.

We should learn to hold emotions like guilt, shame, sorrow, dread, fear, anger, powerlessness, and overwhelm in a healthy and healing way. We should allow our hearts and minds to really process these emotions. There's no need to change these feelings. It's essential to acknowledge and accept them. Negating and turning our backs on them will prevent us from honoring the reality of the situation. The wisdom of the living world is contained within our human bodies, and being with these feelings in our hearts is how we tap into this deep wisdom and fuel our responses.

This emotional process is not about bypassing painful feelings such as despair. It's about feeling them as individuals and as humanity. By acknowledging these emotions and increasing our capacity to hold them, we unshackle our motivations to respond.

Such inner work can facilitate a shift in identity. It can help people move from self-interest to a broader sense of self that encompasses the web of life. It may make us realize that we can speak for the Earth. We can let life work through us. This process can liberate us from the reign of our egos. Opening ourselves widely to this journey can become a powerful touch of grace. It can make us realize we are much more than separate people. By connecting to a larger identity and letting it work through us, we can surprise ourselves with an immense capacity for responding. This capacity is backed by the strength and grace of the forces we act on behalf of.

HEALING THROUGH UNION

We can either continue down the path of fragmentation or switch to one of integration. Fragmentation is driven by a degenerative impulse. In this "me versus other" context, there's increasing dissonance and unease. Fear leads to incoherence and a withdrawal from consciousness. The fear arising from the paradigm of fragmentation manifests in further division and a desire for external control, and eventually it seeks total control. The mindset of separation leads to a closed-mindedness that makes learning and evolving more difficult. This mindset guards the ego and the physical and mental systems it has become accustomed to from perceived threats.

Holistic awareness is the path of greater coherence; this mindset can reorganize and integrate a broader reality. It encourages discernment, nuance, and compassion. This path embodies a balance between the physical and mental, as well as between thinking and feeling. Contemplation and meditation are helpful ways of seeing through the sense of separation. They can bring awareness of our interbeing with all humans and living creatures and help us see our impact and presence beyond our lifespan. The paradigm of integration helps us sit with the complexities inside and around us.

SHADOW WORK

We must shine a light on our individual shadows to transform ourselves. Shadow work is about making the unconscious conscious. Shadows are often parts of us that, at a deep level, we dislike or disidentify with. Our aversion to seeing these parts of ourselves keeps them hidden. Yet these shadows play a prominent role in shaping our behavior. Our shadows can be destructive, especially if we refuse to acknowledge them. They can indirectly call for our attention by making a surprising entrance—seemingly out of nowhere—into our bubble of perception.

Sometimes, these shadows have ancestral origins or are passed down through family or society. The way to deal with them is by cultivating an openness and acceptance of ourselves as we are. Once we can accept ourselves as deeply as possible, including aspects we may find highly problematic, we're able to see these parts of ourselves more clearly. Often, what we dislike or find aversive in others is a good clue to our own shadows.

Aggressively suppressing these parts of ourselves likely makes things worse. It's best to acknowledge them and be compassionate with ourselves. We can seek external help or let our awareness gradually weaken our shadows. Some shadows may take a long time to recede or may never disappear. The main point is learning how they shape our thoughts, feelings, actions, and, ultimately, lives and legacies.

RELATING TO OUR EGO

There's no use in fighting the ego. It's an unwinnable battle. Besides, the ego is there for our survival. But there's no denying that the ego can cause incredible destruction. The way to deal with it is to redeem it. It should be dethroned and relegated to an appropriate position. This is done gradually through self-reflection and awareness.

We have the capacity to compensate for the deficits of our egos. Trying to balance ourselves is a good start. For instance, rather than exemplifying the extremes of being a thinker or a feeler, it's best to take turns and not be a slave to either thoughts or feelings. We can find truth and error in both. Neither our *thinking* nor our *feeling* should be left to decay. Aim for balance.

The ego is malleable and changeable. We can expand our ego, and let it open up. The more elements its sense of "self" encircles and absorbs into its identity, the more wisdom and complexity it can integrate. Contemplating and meditating on concepts such as impermanence, interbeing, and non-discrimination may help keep the ego in check.

KNOW THYSELF

A time-honored path of inner transformation is to turn our attention inward so our body-mind can investigate itself. There are several ways to do this. We can use reason to have an internal conversation about "Who am I?," questioning our answers and inquiring more deeply. That's a good way of finding out what we are not. Self-inquiry may be a good place to start because knowing who we are can help point us in a positive direction. Ideally, self-inquiry will help us arrive at a less rigid concept of self. It's a process of *unlearning* rather than holding on to a specific idea. Another option is to maintain a curious and sincere mental attitude, and to approach the question in a non-verbal way, vaguely conceptualizing the inquiry of "Who am I?" The purpose is not to attain a verbal answer from our mind but to notice something in a new light about our mental chatter and our ordinary experience.

Meditation is another way of learning (or unlearning) about ourselves and our inner workings. One of its initial benefits is the strengthening of our ability to focus our minds' attention. A good initial practice is to sit comfortably in a chair or on the floor (with a straight back and while avoiding falling asleep). Focus your attention on the sensation of your breath as you inhale and exhale. Avoid forcibly controlling the breath; breathe naturally and effortlessly. You can center your attention on your nostrils if you like. Likely, your mind's focus will soon wander off, and you'll think about something else. That's okay. Redirect your attention back to the breath; think of it as the unit of repetition of this exercise. This noticing and redirection is an essential practice. With training, ideally, more time may pass between distractions, and your focus may remain on your breath longer.

Another type of meditation can be practiced by simply observing and recognizing any experiential or mental content within ourselves—without constraining or judging it—including thoughts, feelings, body sensations, and external sense perceptions.[132]

Meditation can help us shift from being caught and absorbed in our thoughts to being better able to let them come and go. This *witnessing* might create enough of a gap to allow us to healthily detach from thoughts and sensations. It's an important step in gaining deeper insights into ourselves and others. This gap will give us a clearer view of the interactions between others and ourselves and between our feelings and thoughts.

These practices increase our awareness of our own dynamics and the relationships between our body-minds and everything else. It's not just a way to improve ourselves; it's a way to know ourselves. Knowing ourselves better benefits everyone because it gives us a more holistic insight into ourselves and our roles as parts of a whole. These practices help us develop right view, right thinking, right speech, and right action. They also declutter our minds of the constructs we've gathered. They allow our minds to relax and open up to insights about and from our unconscious. As we face our crises, these practices may help us avoid frantically rushing into survival mode and operating in the extremes of numbness or apathy, neither of which is helpful. When we practice balance, it helps us discern the "right action."

THE EMBODIED WORLD

Our capacity to redirect our attention from our bodies and senses into our internal mental world has created a false division between mind and body. There is a certain distance between the two, but they are intricately connected and inseparable.

While many of us tend to focus our attention on our minds or away from unpleasant feelings, in the long run, it's best to turn to our bodies and our senses. There's a deep intelligence in our bodies, and turning away from it is turning away from the wisdom gained through millions of years of trial and error.

Connecting to our bodies can help us be with ourselves and find tranquility in simply being. This ability to *simply be* is a healing and nourishing practice. It provides the foundation for empowered action.

Being more embodied allows us to connect with our broad intuition. Our minds are not restricted to the inside of our skulls. They process signals from our world, including the outputs of other minds around us. The nervous system—the mind's substrate—not only extends throughout our bodies but also takes in the signals picked up by our senses. In a very real way, our minds extend beyond our bodies. Many of these signals are processed in ways we're not attuned to. They're mostly unconscious. When we tune in to our bodies, we tap into the wisdom of the body-mind.

Breath work can be a good practice and provides a focal point for feeling our way and playing with the interactions between the body and mind. We hold many energies that shape us, and our actions and breath are portals into the dynamics of these energies.

Practicing embodiment will help some of us notice that we are retaining energies of resentment, judgment, fear, anxiety, anger, sadness, and so on. We may not be aware of these energies, and this awareness is the first step in releasing some of them.

PRACTICING GRATITUDE

The world is a gift. When we forget the gifts that make our lives possible and don't reciprocate and take good care of them, these gifts disappear. Gratitude is not only healing; it's also a reminder of our interdependence. Gratitude and reciprocity are essential principles in nature, and when we disregard them, disastrous consequences follow.

Feeling our interdependence and vulnerability feeds our gratitude, relationships, and humility. When we're overwhelmed by the realization that so much is beyond our control, we can find gratitude by focusing on the present moment. Gratitude is resilience. In the most challenging circumstances, anchoring ourselves in one or two things we're grateful for has a powerful effect on our ability to endure. Ultimately, gratitude is independent of our circumstances. It's an attitude we can practice and improve to draw strength and empowerment from.

LIVING WITH PURPOSE

We're storytelling, meaning-making creatures, and we need stories and meaning because, in some ways, we're too self-aware. We won't transcend that need anytime soon.

Many of us have stopped attending to the old stories and teachings. Some of us have adopted the stance that life has no larger meaning or purpose, that those are just stories. Yet we may forget that this is itself another story—the story of no purpose. However, this story is toxic, and the evidence that it is inaccurate is plain to see if we pay close attention.

Whether we believe that there's a deeper meaning and purpose for us to tune in to or that we must create our own, these perspectives are much closer to the truth. They're much healthier, more generative, and more empowering. Living according to a purpose aligns us better with the evolution of creation, the rollercoaster we call life. It also provides us with a powerful source of resilience. Acting in accordance with our deep expressions is what nourishes our whole being. Life craves both experience and more life. As part of life and co-creators of the world, we should align our unique purpose with this life force. Ideally, working on ourselves, including our meaning, ego, shadows, and so on, allows us to get closer to equanimity and clarity. That's the ideal state of mind for responding to the polycrisis.

Outer Responses

Inner and outer responses go hand in hand. External responses are a sort of "action prayer" that replaces the wishful thinking masquerading as hope.

We can't know our precise individual influence. We are embedded in complex systems and have an unpredictable impact on the world around us. The challenges we face invite us to examine and reshape our personal stance about life and to choose how we want to live and for what. Each of us can reflect on our place within this entanglement and take stock of what we can do to shape our individual and collective futures. Our lives are full of choices; we choose what to encourage and discourage. The navigation aids described earlier can help guide us as we make these choices. Do we wish to embody love, truth, beauty, life, conscious creation, wisdom, and so on? Or do we want to serve the negation of life, the obscuring of reality, and the domination of others? Both paths have undeniably strong forces behind them. Regardless of which one we think is right or wrong, better or worse, these are our choices moment to moment.

TALKING COLLAPSE

Some of us want to talk to our friends and family about the subject of *collapse*. This should be done carefully, gently, and without pressure. Some people may not be psychologically equipped and supported at the moment to engage with this subject and respond wisely. Others may feel significant relief realizing that other people are also aware of these issues—such people greatly benefit from having someone to talk to. Holding this awareness as an isolated individual can be toxic and alienating. When a related subject arises in conversations, we can test the waters to see if the other person is willing to talk about this.

Some people assert that "ignorance is bliss," that the topic of *collapse* is toxic, and that it shouldn't be promoted. That's a narrow view. Denial is a powerful mechanism; if people don't want to hear about it, they will quickly let you know directly or indirectly and retreat to their preferred views.

Many people, especially young people, are willing to engage and respond in light of this awareness because they have more skin in the game and are on the front lines of these issues.

In a world with widespread disinformation, censorship, and narrative control, face-to-face communication between friends and family is an important alternative. The trends of societal decline and deterioration will gradually continue to materialize. As seemingly unrelated issues arise, people will seek explanations. History has shown that people tend to look for simple answers and solutions. Often these simple so-called solutions add to the suffering. That's why our collapse-aware voices matter.

A technique for inviting others to speak about the predicament is to ask them to complete open-ended sentences, such as "When I think about my world, what I'm scared of happening is X," and "As I look at my world, what breaks my heart is Y."[133] This can allow powerful conversations to happen in a way that would never happen on its own. Considering how we ourselves would have liked to learn about this subject can guide us in approaching it with others.

Talking about the gut-wrenching reality of our crises could be the first step towards responding to them. When we share our views, we should do so carefully. Some people believe it's irresponsible to highlight such complex problems without suggesting ways for people to hold their emotions, embody new stories, and respond appropriately. We should offer each other counseling, care, and support. It's also essential to realize that despair can be liberating, allowing some of us to get unstuck and find or create alternative paths.

If you talk about collapse with people unfamiliar with the subject, there will probably be misunderstandings and disillusionment. People generally don't want to know, refuse to believe, and assume you're depressed, irrational, or trying to bring others down.[134] It's crucial that when we have these conversations, we respond warmly to people's emotional reactions. It's good practice to try to relieve any emotional distress and allow people to express their feelings. Expressing empathy helps reduce feelings of isolation, shows some solidarity, and validates thoughts and emotions.

We should speak the truth while balancing it with realistic inspirational sentiments to avoid paralyzing others or leading them to give up their sense of agency. Aim to be precise, gentle, and direct without jargon. It's important to connect genuinely with those listening.

No one should be forced to talk about collapse. Collapse is like a magnifying mirror that compels us to examine our individual and collective shadows and our relationships to death and loss.[135] This type of journey is best walked at our own pace.

Hospicing Modernity

Although the systems supporting civilization are resilient, they are running out of vitality. The timing of the shocks is unpredictable, but the dynamics of diminishing returns have begun. Many aspects of civilization are destructive, and some people are working on hospicing those aspects.

As civilization twists, turns, and writhes, it will cause much harm; for this reason, we should reclaim our sovereignty and hospice harmful aspects of modernity as individuals and as a collective. Personally retreating or withdrawing may be a momentary phase of the path forward. However, we should always strive to return and participate in this unfolding.

PASSIVE HOSPICING

In many places, young people feel disenfranchised by the system. There is increasing dissatisfaction with modern living conditions, and people are signaling this discontent. Some are silent-quitting their participation in the system. Often, these people live a minimalist lifestyle and have given up on the dreams modernity has handed them. They work minimum hours and spend minimal resources on the economy. Instead of committing to the rat race, they are disinvesting most of their energy and efforts while remaining part of the system by necessity. These people are giving up on the ideals and goals of their societies, often because they find them unattainable or not worth the effort. However, this is merely a coping mechanism rather than a deliberate response aimed at shaping our future.

As individuals, we have lots of influence on the quality and quantity of our energy, resources, and attention. Given our crises, we can reduce or redirect the time and energy we spend on things, goals, organizations, institutions, and projects that are not wholesome.

There are many ways in which we can personally divest from and boycott the projects of modernity that are exacerbating this unraveling. Ultimately, the passive approach is about reducing or stopping our support. Since everything is connected, there are myriad ways in which our words, thoughts, and deeds support harmful projects. There are no simple choices that can be prescribed. Each of us lives in a different context, and our responses are unique.

ACTIVE HOSPICING

Active hospicing is more challenging than the passive approach and requires a lot of heart. It exemplifies *acta non verba*: acts, not words. There is often a component of self-sacrifice involved. The insight of interbeing and the sacredness of life provide the foundation for this love-in-action.

An active approach to hospicing the most harmful projects of modernity must be much wiser, more careful, and more skilled than the passive approach. The catch is that every action entails a reaction. For instance, when we use a finger to push an object, there's an equally opposing force between the finger and the object, and there's a third force behind the finger—the arm—which causes the movement.

Active hospicing is often met by an opposing force. Given the dominant narratives and social, economic, and psychological dynamics, such opposing forces tend to be overreactions. This creates a dynamic in which active hospicing performed in unwise and unstrategic manners strengthens rather than weakens the projects and forces being opposed. Even if a particular harmful project falls, more rise in its place, like the heads of a hydra.

How do we grapple with this dynamic as individuals seeking to participate in active hospicing? Overt and direct confrontation will likely be met with an overwhelming force that is more experienced and effective at confrontation. Many social movements have tried to exploit overreactions to their advantage, but these energies forge a life of their own. They follow their own directions, which is not necessarily bad, but is worth noting.

Rather than oppose these projects openly and directly, a holistic approach focuses on severing the roots—reducing the capacity for existing projects to sustain themselves and for new projects to arise. These should be central objectives for active hospicing.

With those considerations in mind, there are nuances to explore: Actions that aren't overt and explicit probably won't give rise to an opposing overreaction as long as they are attributed to unrelated forces and issues. This may be achieved by discreetly "placing sand in the gears." The narratives, energy, and elements supporting these projects are potential targets to undermine.

Another principle to explore is that of a third force. All civilization projects are naturally in opposition or threatened by internal and external forces. For example, a windmill may be threatened by gale-force winds, moisture, rust, sun degradation, design flaws, and a lack of spare parts. All projects have inherent weaknesses, and often their strengths provide clues about them. Active hospicing should align with these weaknesses and forces of decay, because such third forces provide the additional leverage needed to break the balance between hospicing actions and opposing reactions.

A potential trap in active hospicing is that it can reinforce and feed the paradigm of separation, of "us versus them." Since the paradigm of separation is at the root of these predicaments, encouraging it is unwise.

It's vital to hold the perspective of interbeing at the center of any response to this polycrisis, *especially* in active hospicing. Each of us is uniquely positioned to help undermine harmful narratives and projects. We're invited to tune in to our hearts and align our responses with life's divine and sacred force. The life force is expressed through us in unique ways. It's essential to honor and respect the different expressions arising in others, even if we wholeheartedly disagree. Ultimately, the only thing in our power is to be true to the unique manifestation of the force of life in us.

Shaping Our Future

We are co-creating the future from moment to moment. Our everyday lives and personal projects are some ways that we do this. The metacrisis is an open invitation to work on the unlimited areas for improvement. There's no shortage of things to do. The house (i.e., Earth) always wins, and life will prevail; in light of this, we should consider what it means to embody the insight of interbeing.

These challenges invite us to engage in ways that we may have never foreseen. This can represent a significant liberation from expectations that may have been holding us back. There is no limit to how we can respond, which changes everything and nothing. We're still humans with a limited lifespan and also a timeless and unbound potential to make the world a better place in a way that echoes into the future.

Maintaining a balance and not losing perspective when responding to this unraveling is critical. When we lose perspective, it can lead to burnout or increased suffering.

We can get the ball rolling by forming a closer relationship with a small area of land. For instance, we can build a birdbath or take care of a small habitat, like a creek or patch of forest, by cleaning it and improving its health. These humble acts can inspire a strong emotional and spiritual connection to the land.

Any step we take to encourage personal wholeness and social coherence is the right step.[136] Mutual aid will be essential as we weather the coming storms, so developing the foundations for grassroots networks is a great start. Each of us can help facilitate the emergence of initiatives, ranging from community gardens to decentralized infrastructure. Fulfilling our needs for community, shelter, clothing, water, food, and energy in a more localized, self-sufficient, and appropriate way will become increasingly relevant.

Individual Preparedness and Resilience

The challenges of the polycrisis will increasingly impact us at both the individual and collective levels. We shouldn't disregard our individual resilience in favor of the collective. Each of us is in charge of ourselves. We should think and act in ways that enhance our resilience as individuals and families, because the less prepared we are, the harder it will be to withstand the shocks that come our way. Getting caught with our pants down would make things more difficult at all levels.

Being prepared has gotten a bad reputation in many places because it's perceived as an individualistic and selfish response. People interested in preparedness are often portrayed as gun nuts hoarding stockpiles of goods, eager to isolate themselves and deny help to others. Of course, some people take things to the extreme.

But those who view individual preparedness negatively may be projecting their own shadows of selfishness, insecurity, and impotence onto those who prepare. After all, it may be triggering to ponder how helpless one might be in the event of a widespread, long-term disruption of goods and services.

Often, critics of individual preparedness don't view the readiness of their governments in the same negative light. Many governments have stockpiles, ranging from fuel to medical supplies to food, and practice some level of preparedness, even if their capacities are insufficient. These governments prepare at a national level, not globally, so it can be seen as "self-centered" behavior—but it's not; it's reasonable and rational.

Increasing our resilience and preparedness as individuals or families is a good idea. The less resilient we are as individuals, the less resilient the collective is. In the emerging world, there's no doubt that individual preparedness will play a helpful role. Governments can't be the sole source of resilience, and they never will.

The members of a tribe not only rely on the whole tribe for resilience, but their bodies also store supplies—fat, protein, vitamins, and minerals—in case they face challenges. Within our bodies, there is redundancy and resilience in having two sets of certain organs and in many other natural processes. There are important reasons why nature has given us resilience at multiple levels, as individuals and at levels nested within our bodies. To bash individual preparedness is to reject the wisdom of nature.

As a starting point, consider planning for a situation where food becomes too expensive and we are forced to rely on food banks or government rations for some of our sustenance. That's the current reality for many people worldwide. In this context, it makes sense to grow some of our food. Living communally or in a mutual aid community is also a good way to share resources and work together.

Having Children in Unprecedented Times

The decision to have children can be viewed from both individual and collective perspectives. Some people are concerned about the future well-being of children, or feel guilty about having a child that contributes to the crises, or both.

From the perspective of civilization's impacts and resource requirements, there's already an overabundance of humans, and we are in ecological overshoot. For instance, wildlife populations have collapsed as the human population has increased exponentially. For every civilization dweller added, there's a roughly proportional subtraction in wildlife. We are also negatively impacting the living world's capacity to sustain our population, given our numbers and rates of consumption. From a collective angle, increasing our population makes things more challenging for humanity and for other living beings. The current global fertility rate of children per woman is 2.3, and to reach stability rather than growth (all things being equal), it has to decrease to 2.1.[137] Having two children or fewer per woman is a sensible suggestion. Regardless, the human population will probably peak in the mid-century.

Ideally, the decision to have children or not should be personal, and it should also consider the well-being of the community. Having none, one, or two makes more sense, considering our predicaments. At the same time, it's not just about reason; a decision like this should also be about honoring our deep and true impulses—it shouldn't be a calculation coming from the mindset of separation. We live in unprecedented times, and it's good to be mindful of the collective and individual context we find ourselves in as we consider having children. We will be tested in many ways for the foreseeable future, and holding this awareness as we make decisions will help us be at peace with whatever we choose.

Many couples pondering the decision may feel uneasiness, anxiety, and fear of an unknown future. We should remember that we're the continuation of a lineage that extends back to the beginning of human history—and beyond—and has endured incredibly tough situations. History is bursting with stories of adversity and endurance. If emotions of wariness about the future are involved in the decision, I suggest considering the difficulties of our collective past. Parents have never been able to guarantee their children's safety or reassure them that everything will be all right, and they never will. No one can guarantee what the future will be like. Having a child is always an incalculable choice.

A child who understands that humanity has a place and a purpose in this world has the potential to be a genuine gift to the Earth; anyone who embodies nature's life force and a sense of interbeing can be such a gift.

Several aspects of life may become more challenging in the future; likewise, different aspects may be less challenging. Our era offers us an extraordinary opportunity to live a life filled with meaning and purpose and to contribute to something much bigger than our individual lives. A lack of adversity is not what leads to happiness, after all.

As individuals and couples, the decision to have children should ideally be made with both heart and mind. Each of us has our own unique context and responses. This topic touches on our deeply personal philosophies about life's meaning, facing suffering, and what makes a life worth living. It touches on our intimate worldviews. Those pushing their opinions onto others are projecting their personal feelings and rationalizations.

Responses to Be Careful About or Avoid

The dominant narratives don't offer much helpful guidance for facing this unfolding. Thus, many succumb to nihilism or direct their upset inward from the lack of direction or a solid foundation. The sudden disintegration of our default worldview can create a vacuum in personal meaning, but nihilism is an incorrect view. Discovering or creating our own meaning is much more empowering, coherent, and healthy for humanity and life.

Nihilism can result in apathy and selfish hedonism. Both encourage division and lead to individual and collective suffering. Those who "check out" from the collective and don't care about what happens outside their bubble enter a cocoon of self-delusion. This cocoon can only provide temporary relief; reality will eventually pierce through, demonstrating harshly and overtly how we are all interconnected. Likewise, pursuing hedonism is not sustainable—due to the coincidence of opposites, significant pain is bound to catch up.

Partying until the music stops is the current collective dynamic. It's a self-centered response that encourages self-destruction instead of improving our situation. Those of us on the nihilistic path are turning our backs on our ancestors' struggles, our fellow humans and living beings, and future generations, and saying, "Nah, I don't care."

If we're fully aware but don't respond in alignment with life, then deep in our hearts and minds, there could be an intuition that we're avoiding something. We won't ever be completely engaged with reality.

Directing our upset inward often leads to internalized guilt, judgment, or disapproval. These energies are generally directed at humanity and sometimes lead to misanthropy. Seeing humans solely as having a negative impact or as flawed beings unable to stop waging war on each other and nature can become a self-fulfilling prophecy.

If we reduce humans to narrow variables of environmental impact—like carbon emissions—we could rationalize that humans have a net negative effect. That view is problematic. Such a train of thought can also lead us to believe that every human action could be analyzed through a few limited variables. However, reality can't be reduced to a set of variables, nor can the impacts of humanity or individual humans. When this reductionist mindset is taken to the extreme, it can lead people to develop a false sense of righteousness that justifies their crusading, judging, and shaming of each other. It's a counterproductive approach that paints the world in black and white and strengthens our collective obstacles.

Taken to its limits, a reductive approach fixated on reducing harm can make people feel bad about their lives. It's disempowering and can drive people and societies to a toxic state.

The patterns of urgency and haste aren't solving any problems. Activism and community organizing have been in a "resistance and demand" mode for decades. They have focused on resisting bad ideas and demanding change from the "people in power." The problem is that this routine reinforces the hierarchical power dynamics, and the establishment has repeatedly made it clear they are not willing to redistribute that power.

So we have to exercise our own power, not for the sake of power or control, but to assert who we are. Somehow, we must find ways to exercise our community autonomy, take things into our own hands, and claim our power. We should go beyond bargaining for crumbs of incrementalism, for doing so entraps us in a victim-slave mindset.

Escapism

There is no escape. Eight billion humans won't escape their issues by going to space. Our planet needs no terraforming, no cosmic radiation shielding, and no industrial production of oxygen or water. It needs no soil building and requires no multi-month travel time or lengthy communication delays. It demands no complex infrastructure. We are on board the best, most resilient, and most spacious "spaceship" we could ever build. This is it. Hopefully the space-colony dreamers will slowly realize that what they've been looking for is here.

The visions of propagating life outside our planet are an escapist dream, neglecting the immediate challenges. This dream is completely inconsistent from a life-affirming standpoint, because the same people pushing these ideas are disregarding the current crises related to life on our own planet. On one hand, they believe humanity doesn't need the living world for survival, and at the same time, they use the story of "propagating life throughout the universe" to justify their delusions.

Aggressive Resistance

Aggressively resisting the dominant enterprise of civilization will probably strengthen the resolve of the status quo and the denial of the issues. Many decision-makers, rulers, and those who promote mainstream narratives are driven by status, fear, power, and a need for control. The metacrisis shows that the emperor has no clothes. It exposes some of the "drivers" behind our collective insanity: *insecurities* and an *addiction to control*. Many of these people find or might find themselves caught between two sides. If that happens, they will likely side with the habitual paradigm of business as usual. Most of them probably have no other choice due to their conditioning, circumstances, and the pressures behind them. However, they should be offered ways out. These people should not be cornered into seemingly inescapable positions; if they are, they will probably redouble their resolve and fight like their lives depend on it.

When we frame our situation as a war and identify an enemy, we oversimplify the problem and turn it into something that modernity is quite experienced in dealing with. In this way, we also deceive ourselves about the nature of the issues and their solutions. If there's an activity at which modernity reigns supreme, it's waging war.

Everyone taking part in civilization is partly responsible for this metacrisis. Blaming one group of people or another will encourage division. It's not just a matter of strategy; blaming an external group is self-deception and minimizes our own responsibility and agency.

As we make a stand, we should minimize the fear, violence, and division we propagate. This is not easy. However, anger and violence certainly have a place in this world. There's a reason we have such capacities. It's up to us to honor our deep impulses with great discernment and wisdom, because they can ignite powerfully and spread quickly. These energies can become quite toxic and forge unpredictable paths. They are not solid foundations on which to ground ourselves for the long run; they have a shallowness rooted in fear and ignorance.

It doesn't matter if our actions won't prevent a collective *reckoning*. What matters is that we work with what we've got and try to make the most of it. As the polycrisis unfolds, triggering deep-seated fears and unease with our actions and cornering people into inescapable situations is a sure way of making things worse. Resistance has its place, but if all we do is resist, we'll learn to love resistance. We may have plenty to fight against, but nothing to fight for.

We should focus on meeting our needs in alignment with life and the insight of interbeing. When we meet our needs as directly as possible, we form a strong bond of interdependence with life and the land. In this way, we can defend against and resist any infringement on our sources of life and sustenance. We can tap into a stronger, more grounded energy as we exercise our resistance and sovereignty. For our efforts to maintain their vitality and develop strong roots, they must be founded on interdependence.

We should form roots in the land, water, plants, and animals around us. A healthy community aligned with life can be a foundation on which to stand as we respond. By acting from a place of connection to the land, we will learn how to sustain our actions in the world at large.

Those at the front lines of the struggle against the destruction of the sacred should assert their power and sovereignty. They should mediate their relationships with the land and living world on their own terms. Being in solidarity with these people is not about sympathy or altruism, but a recognition that our well-being is fundamentally interconnected.

11. Collective Responses

This chapter explores some broader collective responses that may emerge as the polycrisis unfolds. We shouldn't passively count on these responses arising. Some of us will be called to work on them. Still, this book is not the type to prescribe a twelve step "solution" to the issues. Instead, it shares potential paths to explore.

Personal vs. Collective

Often, advice on changing the world encourages us to work in groups rather than individually, which is how change scales up. However, change has always had both individual and collective components. For instance, an individual may develop a new process, idea, or invention that has a significant impact on humanity and the world. But without a collective component taking up that initial spark, change wouldn't happen.

Each of us has something unique to contribute. Most of our individual impacts are not the result of grand gestures or being part of a group, but rather stem from the lifelong accumulation of our trivial and non-trivial personal choices and behaviors. Some people may need time and a fertile environment to forge their unique contributions, and joining a group could be distracting for them.

Of course, joining with others can be an empowering way of co-creating change at scale. But it's good to be discerning, because groups and organizations can capture our energy, time, and passion and use them for their own goals. As organizations mature, they're prone to redirecting their efforts from foundational goals to self-preservation.

It's crucial to collaborate with others to have a broader impact. But always be aware of the relationship between your personal vision, efforts, and spark and that of the group or organization you belong to. When joining an organization, consider its needs and how it meets them. Follow the money. Who benefits the most, directly and indirectly? What are the organization's biases, limitations, and outcomes?

Many existing organizations have such great inertia, structural constraints, or limitations that they cannot effectively respond to the polycrisis. They may pretend to consider cherry-picked implications that are not too inconvenient or contradictory to their agenda. However, many of these legacy organizations are practically impotent in response to the polycrisis. Even worse, some turn us into useful idiots who help rich people make a bunch of money without tackling any root causes and while keeping existing power dynamics intact. These organizations are a massive drain on our energies and efforts. It's crucial to assess whether the group we're joining has internal contradictions that render it incapable of responding adequately.

Suppose we already find ourselves embedded in an organization. In that case, we should consider the *real* impact we can have there, regardless of the kind of organization it is or how seemingly unrelated its mission is to the metacrisis. It's best if people who understand the collective situation are embedded in all sorts of organizations.

Societal Responses

Acknowledgement

Realizing what is happening is a reasonable step towards responding to our challenges. Collectively, we will be better able to respond if we go beyond denial and enter into acknowledgment. Humanity has a strong capacity to share messages quickly and reach almost every person on Earth. We have a big issue, so we'd better talk about it.

Those suppressing the conversation don't know what they're doing or who they're serving. They are free to disagree with the notion that civilization is in a severe predicament, but suppressing this information only makes the predicament more likely to escalate and more challenging to address. A person may have a terminal illness without being aware of it. No one may know precisely the timeline or if the patient will die from it, but not talking about the illness's existence or severity is not only unhelpful; it prevents timely and appropriate responses.

We're in an environment where many bottlenecks slow down the flow of information. In addition to the natural barriers of the status quo narratives, social media ownership, cancel culture, and censorship make it difficult for ideas to breach the normalcy bias bubble. When all of that is coupled with our psychological tendencies, it's clear that spreading awareness of the situation is challenging.

Modernity has huge shadows that remain out of sight for most. To understand our metacrisis, we must shed light on these collective shadows.

We must experience collective grief to honor this moment. We require such a psychological and emotional collective reckoning if we hope to enlist all our resources as human beings. We can't just hold this awareness from a distance. We won't be able to tap into our deeper, wiser, and most potent impulses unless our hearts are fully engaged and feel the full intensity of this moment.

Alternative Narratives and Cultures

We need stories to make sense of our existence, the world, and the arc of our lives—from birth to death. The stories of our time and those of our parents' time shape our lives. They give us our hopes, values, and beliefs. They shape the institutions and systems around us.

However, the current dominant myths and stories do not serve us. We're in service to them, and they're taking us down the path of negation, of self-annihilation. The old stories are failing us—the proof is in our predicament. We direly need new stories.

Ultimately, our cultures and societies must align with life if they are to achieve longevity; there is no alternative. For some time, we have sought meaning in the stories of modernity, but it's now our turn to express, as representatives of life, what we want life and creation to be about. The universe is questioning us, and we no longer need to ask outside ourselves about the meaning of life. These are the times for us—as life itself—to show through thought, speech, and action why we're here.

We are free to acknowledge and honor the parts of the old stories that worked for us, but we should stop clinging to the ones that no longer serve. These latter narratives are proving to be tremendous obstacles. Our societies are undoubtedly teetering amid a struggle between competing narratives. The paradigm of separation seems to be firmly entrenched. Yet gradually, the paradigm of interbeing is finding spaces to flourish in.

For the new stories to possess a subversive force, they should be upsetting and uncomfortable for the dominant ones. The new stories should shed light on both our individual and collective shadows. This is a significant opportunity to shape our cultures. We're letting go of the old stories of self, society, and the world and seeing the emergence of new ones. These new stories can expose and undermine the narratives and systems that divide us.

Not only do we need new meanings and values that align with life, but we also need new ways to satisfy those meanings and values. This area has great potential: Stories and narratives can move mountains. They can create shared interests, identities, and ideals. A simple, vague, but effective myth that provides a sense of belonging and a collective "self" could serve as a guide and as the glue that facilitates the emergence of a collective shift.

As we strive to adapt, we will undergo tremendous changes. It won't be a matter of symbolic gestures. Our societies will be greatly transformed, and narratives can support and guide our efforts as we adapt.

Using reason as the main guide and engaging humanity as a *collective* in the new stories and myths won't be powerful or effective enough. The new stories must be deeply personal. They must touch our very core. To achieve this, artists creating these visions should turn inward to genuinely connect with others.[138] The foundation of our collective future is seeded at an individual level.

Human cultures have faced apocalyptic times before, and they have endured them through ritual practice and a deep trust based on interconnectedness.[139] This trust goes beyond communal relationships. It's an embrace of the core forces that shape the universe. This deep trust in mysterious wisdom can provide us with an unshakable foundation on which to stand as we journey on.

Stories can be a fantastic aid. They can reshape how we perceive ourselves and our roles in the world. They can help us remember that we're not only *takers* but can also be *givers* and *partners*. They show us the good things we can bring to the table.

We must get back in touch with our mortality, sorrow, and gratitude for life. Stories, art, and rituals help us achieve this collectively. We must build our capacity to engage with deep grief and despair and understand that they don't have to be fully immobilizing. Our ancestors have dealt with these emotions before—we can handle them too.

Cooperation and Mutual Aid

The predicaments won't dictate how challenging our future will be; our collective responses will. Any appropriate real-world response will have components of both competition and cooperation. However, a response based primarily on cooperation will be the most likely to succeed—eons of evolution back this. We wouldn't exist if life were primarily a competition of all against all.

This is the mother of all crises. If there was ever a need for humanity to come together, this is it. To be clear, the geopolitical world will fragment; there's no doubt about it. But the more mutual aid and solidarity there is, the better our future will be. For instance, collectivizing resources (at a grassroots level) is an approach that nurtures social cohesion and agency, and provides needed goods and services in a resilient way.

Resilience

The evidence suggests that modernity, as a complex system, is somewhere between the *conservation* and *release* stages of the adaptive cycle. If a shift into the release stage this century is likely, what can we do?

No single person can speak for life or reality. One life-aligned response would be to strive for conditions that allow the continuation of humanity and a healthy living world. If we care about minimizing human suffering, we would encourage conditions that enable us to better withstand the shocks that will force civilization to simplify.

The mainstream school of thought on resilience focuses on keeping civilization, the system of systems, functioning as it navigates the polycrisis. The hidden premise is that humanity is utterly dependent on industrial modernity. Civilization is deemed too big to fail. There's a problem with that. Complex systems behave in a cyclical pattern of growth and collapse. It's the intrinsic nature of complex systems, including all living beings (and their communities). It's an inescapable reality. Civilization won't be bailed out. No amount of make-believe will help. However, we can't be sure when it will unravel.

A wiser and better-informed school of thought on resilience would focus on keeping humans and the living world as healthy as possible as we navigate the predicaments of civilization. There are some distinctions to make about this approach. The usual approach focuses on perpetuating a self-terminating, lifeless system without regard for nature or interbeing. A resilient one focuses on maximizing the well-being of life and minimizing suffering. These are not the same. True resilience is oriented to life, not to a system.

For all its complexity, life is quite simple. Humans need a healthy living world, community, water, food, clothing, shelter, safety, medical care, and means to cool or heat our shelters and cook. These needs must be met in a manner that is balanced with nature. Beyond that, they're not *needs*. Genuine human-focused resilience would concentrate on those simple needs.

Additionally, one of the best ways to enhance our resilience at the ground level is by expanding our social capital. Some examples of this are close-knit communities where neighbors help each other or rely on family members for support.

Activism

There continues to be a role for activist initiatives. We should persist in expressing our concerns and working on our visions. Dissenting voices will undoubtedly be pressured to find alternative means of grabbing attention if the establishment refuses to hear them. Direct action, civil disobedience, and protests send far-reaching signals to people, organizations, businesses, and governments.

Those involved in activism should carefully consider what's being signaled through their actions. Are they spreading fear? A sense of superiority? Is it virtue signaling? A threatening message? A supplication implying powerlessness? Is the message aligned with life, or with something else?

The medium is often the message; we should attempt to understand the symbolism and dynamics we propagate. Are the actions communicating solidarity? A sense of deep love and compassion? A sense of responsibility and agency? A profound conviction? An unstoppable force? The will of the world? Ideally, the energies propagated should be aligned with transcendent forces and insights rather than with our personal and collective egos or shadows.

Activists should carefully consider the messages they convey through their activism, words, and actions. What narratives do they feed? Who benefits? What future does it serve? This inquiry must be *brutally* honest. Deviating from the truth, even minimally, is not the right way to inquire. It's best to avoid initiatives that focus on incremental reformism and propping up self-destructive systems.

Direct Action

Direct action is not about signaling or making demands but about getting things done—about taking matters into one's own hands. It carries greater risk of backfiring. So it must be done carefully and thoughtfully, and it should be guided by good navigation aids, such as those in the previous chapter.

Recently, there has been an increased criminalization of dissent in the Global North, for instance, by giving disproportionately long prison sentences to activists. As the risks grow, activists will naturally seek higher rewards by focusing on non-symbolic, anonymous direct actions, such as targeted sabotage. The increasingly suppressive context will lead to a transition from virtue signaling and "martyrdom" to campaigns of selfless, thankless, and unclaimed actions made for their own sake.

Organized Resistance

Regarding the effectiveness of non-violence, it's worth noting that non-violent movements often succeed only in the presence of a real threat of violence from other resistance groups. Historically, rulers have probably conceded to non-violent resistance because they would rather not face the threat of violent resistance. Strict proponents of non-violence have often overlooked this synergy between non-violent and violent resistance.

Wars and coups represent some violent means that have achieved change through strategy or brute force, but they, without a doubt, perpetuate the paradigm of separation and domination, and would fail to address the root issues of the polycrisis.

Historically, people have shaped their societies by applying pressure on strategic leverage points in their societies' social, economic, or operational systems. These actions have included demonstrations, debt resistance, tax resistance, boycotts, and general strikes.

Those resistance tactics have variously worked for both small changes championed by a few people and for significant changes advocated for by a majority. However, they may not apply to the current metacrisis, because it entails enormous changes advocated for by a small portion of the populace.

Relatively few people have a sufficient understanding of our predicament. Most people in our societies see modern issues as separate, temporary crises with straightforward solutions. Roughly, their understanding is that they can have their cake and eat it too. However, that paradigm is changing gradually. For these reasons, any tactics that require the implicit support of a substantial number of people don't seem relevant at this time. The most relevant actions don't require large numbers of people and don't need implicit public support. In addition, acknowledgement and awareness may be more important at this stage. Our collective mindset must change for that change to be reflected in our world.

Collective Action

Our lack of collective power and agency is a major obstacle to regenerative adaptation.[140] To reconnect with that collective power and agency, we must build effective platforms for coordination, and more importantly, we must facilitate the emergence of systems that can provide for our real needs, such as food, water, energy, and a sense of belonging and community. That's how power is built. There is no collective power without the ability to coordinate resources to meet our collective needs.

As the metacrisis unfolds, opportunities for action will arise, so we should build positions of strength from which to seize them. Collective change doesn't happen because of key people or groups; shifts emerge from the interaction of all the agents involved in the superorganism. As in any complex system, the tension and interplay ultimately decide the outcome. It's important to note that the strategies arising from collective narratives or groups of groups are emergent. A holistic approach to social change sees personal transformation, the building of alternatives, reformism, structure-based organizing, direct action, and protests as an ecology of complementary, converging actions towards an overarching goal. These diverse tactics and strategies shouldn't compete—they should reinforce each other. All of our approaches should be layered and nested in a fractal arrangement.

A possible collective goal could be facilitating a movement that can support people in meeting their needs, take a stand against state repression, and become a source of resilience as the simplification unfolds. For that to emerge, individuals and groups must align along shared goals and interests, and deliberate coordination must happen at all levels. Coordination could take place on the basis of location or through organizations and alliances.

A difficult challenge with collective action is that our reliance on the status quo blocks us from effectively shaping the systems of that same status quo. Groups seeking social change have been stuck for a long time in a self-defeating strategy of feeding harmful systems with resources and support because they rely on those systems to meet their material needs. Without redirecting resources, energy, and efforts away from modernity and into alternatives, we'll stay on the path of self-destruction. That's why it's crucial to shift to life-affirming cultures and social structures and build the practical infrastructure to support our collective well-being.

Organizing platforms could be established to facilitate the emergence of collective change. These platforms could serve as foundations for the emergence of social shifts by providing sufficient constraints and direction without being too restrictive. Organizing platforms share *pre-established goals* or narratives; *protocols*, such as guardrails for actions, risk levels, messaging, and commitment; and *resources*, such as group identity, infrastructure, aid, and information. The pre-establishment of guiding goals, values, and overarching strategy is an essential initial step as such platforms develop. These platforms can accelerate the development of alternatives and provide space for experimentation and replication of grassroots initiatives. They should always be guided by deep, life-aligned principles. Organizing platforms can serve as melting pots where people and groups connect and invite others to join. They should also aim to meet people's immediate needs and focus on concrete challenges and initiatives. Ideally, they should encourage opportunities for pivoting and transforming as the conditions change, aiming to be adaptable and allow for new strategies and directions to emerge.

To move towards societies that are content with meeting their needs, not their wants, we must share resources communally and use appropriate technologies. These societies must be organized at a small scale by local people and organizations; due to the reduction in complexity brought by the unraveling, these units would have to form networks for large-scale coordination.

This approach to collective action should work at small scales to bring about change here and now while also building collective power and agency for larger long-term shifts in the future. The strategies should be encompassing and have a broad field of participants in mind. They should support people as they develop a sense of belonging, tap into their own power, and find purpose in serving others. Power must be built wherever possible, in the cracks and underground. Developing the capacity for defense from state repression will be increasingly important as a broader foundation is built.

Taking Refuge

As the paradigm of separation and what has been built on it decay, regenerative alternatives will flourish. The people and communities embodying these alternatives won't remain unscathed as modernity declines. But these alternative ways of living and being will nonetheless be helpful as people navigate the simplification.

There will be efforts to form autonomous zones that strive to be self-determined and remain free of state and market forces.[141] These zones might be regional, village-sized, or community-sized. Ideally, people in these zones would embody alternative narratives and paradigms that are socially balanced and aligned with nature. These zones should strive for self-sufficiency in food production, water, energy, and so on. For now, these zones will have to keep one foot in the alternatives and one in the status quo. But they should strive to stand on their own, outside the state and the market.

Supporting the self-determination of Indigenous communities is a no-brainer; many people in these communities have not lost touch with the reality of interbeing with nature and have preserved ancestral knowledge, cultures, and lifestyles.

Techno-utopian narratives paint a picture of futuristic cities that house most humans while preserving the surrounding environment intact, but these visions are contrary to reality. The trend of greater urbanization is inextricably linked to increased environmental damage. The story of "sustainable eco-friendly megacities" is a make-believe tale that doesn't consider the root causes of the crises or the whole picture.

As the trend of urbanization grows, there will also be waves of people moving from cities to rural areas, trying to escape the declining urban landscapes. This can be an opportunity to form intentional communities that embody alternative paradigms and lifestyles, striving for localization, self-sufficiency, and alignment with nature.

Composting

Many parts of civilization should be composted into something genuinely aligned with humanity and life. This applies to ideas, organizations, institutions, infrastructure, and so on. Almost every aspect of modernity aligns with the assumptions of the paradigm of separation: that there are no limits and that the negative consequences of civilization are negligible. Our societies will undergo major overhauls in how they operate. Rather than discarding all aspects of civilization, we can transform what can be repurposed to truly benefit us and the living world.

Appropriate Technology and Infrastructure

The conventional approaches to the crises through the mobilization of technology or infrastructure assume that the past trends in economic growth, technological complexity, and geopolitical and supply-chain stability will continue as before. The visions behind these responses imagine a future techno-utopian civilization powered by solar panels, wind turbines, batteries, and nuclear energy. Some visions foresee space colonies and even mining minerals outside Earth in the near future. These visions suffer from many significant issues, particularly with feasibility and consequences.

In no way does this type of vision hold life as a central sacred guide, and that's a big problem. It implies we don't need life: We don't need butterflies, dolphins, rivers, and forests. We just need solar panels and nuclear reactors, and we'll eat lab-grown food and live inside our environment-controlled bubble-like complexes. This narrative is a distancing and turning away from life. It's a rejection of impermanence, interbeing, our lack of control, life's "messiness" and unpredictability, humanity's place in the living world, and the spiral and ever-changing nature of existence. We don't make the rules of this kingdom, and disregarding them comes at everyone's expense.

Rather than working on the vision of a delusional techno-utopia, responses involving technology and infrastructure should be mindful of and embrace the fullness of the implications and challenges that come with life, and should be planned and considered holistically. They should consider the narratives and power dynamics they encourage, possible unintended consequences, whether they promote social cohesion, and if there are wiser alternatives for achieving similar outcomes.

Considering resilience and viability, appropriate infrastructure and technology should be low-cost, be co-developed by those using it and other stakeholders, have minimal reliance on international supply chains, use minimal resources cradle to cradle, have a minimal environmental impact (assessed holistically), and not be manufactured or controlled in centralized ways.

External Responses

Biological Integrity

Maintaining and supporting the health of the living Earth makes complete sense and has no major downsides. The increasing pressures exerted on the living world, including habitat loss, pollution, species loss, and climate change, will be reflected back on humanity. Mitigating these pressures won't only reduce our challenges, but will help the living world withstand the shocks already baked in.

We must encourage, not disrupt, nature's capacity to adapt. Maintaining the integrity of the living world and preserving other-than-human life are *right action*. However, these efforts should be made wisely and respectfully, in solidarity with Indigenous communities. We should see ourselves as a part of nature, not apart from it.

Mitigation

We need to mitigate impacts to the living world, especially through those opportunities that are low-hanging fruit. Every added bit of damage to the living world harms us and future generations. Understanding that humanity won't be able to reverse all the damage is not a justification for giving up on mitigation. We can do things because they are the right thing to do, because they benefit our societies and our world, or because we find enjoyment or satisfaction in them, even if our efforts won't guarantee a given outcome.

Some thought leaders try to suppress ideas to manage others' awareness of issues like climate change, thinking that people would give up if they believed we were "losing the fight." The mental projections of these thought leaders show that they haven't arrived at a place where they themselves wouldn't give up in the face of overwhelming odds. Clearly, their conviction hasn't been tested; it's precarious.

Mitigation should be done with equanimity, as a form of active prayer. We won't "save the world" or civilization through mitigation. That's not how it works.

We should be honest about whether we're doing things for our benefit—in a narrow sense—or because they're the right thing to do and benefit us in a broad sense. The difference is crucial, because it's the foundation on which efforts are built. Nature has an incredible capacity for regeneration, and any effort we make to help nature regenerate goes a long way.

Responses to Be Careful About

The Role of the State

Nation-states cannot reverse the polycrisis any more than beauty creams can reverse the aging process. However, government responses can influence our capacity to adapt and can either increase or decrease the suffering humans and other-than-humans will experience. Governments should understand their role in this way.

Unfortunately, it's more likely that governments will act under the view that these are problems with solutions, and there's no chance of responding appropriately without a grasp on the complex reality of this unraveling. In addition, many governments will focus solely on staying in power, which will become increasingly difficult and result in a misallocation of efforts.

Ideally, governments would realize what's happening and what's likely to happen—scores of people in governmental positions must have undoubtedly connected the dots already. A wise government response would be to create conditions that encourage the public to draw their own conclusions. They would support independent research efforts and provide transparent and accurate information on the state of the living world, the functioning and needs of civilization, and the negative consequences of modernity. This is already happening to an extent, but it needs to be an earnest, concerted effort.

No doubt there would be differing views and ongoing debates. Yet with increased public awareness, governments would have a better chance of overtly working on influencing the timing, abruptness, and depth of the simplifications. They could cushion and distribute the impacts more fairly, responding with wiser triaging, and could act in solidarity with other countries. This is possible in theory—a version of this is already happening, but not at the scale needed. Attempting this would require selfless, honest, and wise government leadership. That's rare in the current political realm. Governments will likely continue to exemplify ignorance and play a game of hot potato, passing the responsibility to the next in line.

In theory, governments could significantly impact the unfolding; they could do a lot to support the resilience of people and communities and encourage society to transform itself proactively rather than being dragged along by fate.

Minimizing Resource Use

One appropriate response to the polycrisis would be minimizing our overall use of essential resources. Globally and per person, the use of resources continues to increase; a clear example is the rising total energy use worldwide.[142]

We'll have to do some rationing sooner or later. However, it's crucial to keep in mind that mandated rationing and controlling the flow of essential resources have historically led to terrible suffering through issues like starvation and the degradation of living conditions. Rationing must happen in an organic, grassroots way, or it will exacerbate the challenges. Organizing rationing in a top-down manner will almost inevitably screw things up. It would be like allowing our conscious minds to micromanage how much food each cell in our bodies gets. Top-down rationing would be catastrophic. Appropriate rationing would need to be organized from the bottom up. It's the only way to do it without causing widespread negative consequences.

Along with minimizing resource use, there should be a triage to allocate resources appropriately. So many resources are wasted, and many more are used for things that are a relative waste. There's so much waste globally that we can substantially reduce our use of resources while still meeting our genuine needs. This triage should be done carefully, gradually, holistically, and from the bottom up. The actual, genuine needs of people should be met.

Those who cling to the unfounded hope of salvation without adversity are oblivious to the reality that there have been no successful efforts to face the challenges head-on. The fact that collectively we haven't been intentional about rationing, minimizing resource use, and triaging speaks loudly about how assertive and appropriate our responses to this polycrisis have been. So far, the responses have been fifty years' worth of mostly posturing and symbolism.

Counterproductive Responses

Modernity is experiencing a period of accelerated drawdown of resources. Due to economic slowdown, geopolitical shifts, and a growth-based paradigm, civilization is doubling down on resource extraction and industrialization, rather than exercising self-restraint and redesigning the entire endeavor.

This path is not one chosen from the ground up. It's emerging from the interaction between the top and the bottom. Centralized systems are leading us towards a "managed" self-destruction nightmare. The systems in which we are embedded are mostly centralized, and the structure and limitations of these systems are leading us to a different *sameness*. These systems are built on separation, domination, control, and one-way vertical violence.

Feeding a system committed to self-destruction with alternative sources of sustenance will not address the root causes of the metacrisis. Accounting tricks won't solve anything. Getting together to pretend we're tackling these issues won't solve anything either. Fear- and profit-based initiatives won't address the root causes. Running risky experiments on a large scale with our living Earth won't cure the polycrisis. These false paths lead to complacency, enabling further suffering and harm. Many people deeply involved with these issues are not facing our crises with total honesty towards themselves and towards others. Many initiatives championed by the establishment are self-enriching, hypocritical, authoritarian, misguided, narrowly focused, unfair, ineffective, and unviable.

Suppressing public awareness of the full extent and roots of the predicament is counterproductive. These issues are beyond complex, and we can't adequately respond if we lack the humility and equanimity necessary to perceive the truth. Responding based on half-truths or half-lies will probably lead us to where we don't want to be.

This goes well beyond our personal ideas. Either we face the truth as it is, whatever it may be, or we let ourselves be deceived by our own unconscious agendas and biases. Now is a time for embodying true maturity, not focusing on our careers, dreams, status, network, prospects, and so on. It's the time for us to get out of our own way for our own good and the good of the living world.

Authoritarianism and Centralization

Some may argue that social structures based on pyramid-like hierarchies are effective at dealing with crises. A hierarchical pyramid is the easiest way to connect everyone to a central figure or group, but it's also the dumbest way. There's a reason why examples of this kind of organizational structure are rare in nature. This structure can't handle the vast, complex information coming from all the agents in the pyramid during dynamic situations. It's rigid and vulnerable in contexts of massive change and uncertainty.

Many may awaken from the illusion of control as the status quo deteriorates. Many people, finding themselves without much control while trapped in the paradigm of separation, will try to fulfill their desires by projecting control onto central figures. Amidst these challenges, many people will partly regress emotionally and psychologically: "Nanny or Daddy state will take care of me and guide me through this; they know best."

In times of fast-paced change, centralizing control at a national or international level may seem attractive to many, especially to those who currently make the rules. These are the times in which authoritarianism and totalitarianism make a go at it. Populist leaders will try to take advantage of the discontent. The ruling groups may blame issues on scapegoats or lash out against the masses through brutal coercion. In the minds of these leaders and their supporters, consolidating and concentrating power is a reasonable and necessary step towards tackling these issues effectively.

However, centralization is a regression in organizational complexity. It entails a significant decrease in adaptability and resilience. Concentrating control amongst fewer individuals amplifies the negative consequences of their blind spots, tunnel vision, and silos of information. Like a general waging a losing war telling a dictator only what they want to hear, the discrepancy between reality and the information reaching the inner circle's bubble would only worsen the polycrisis. Centralized control would make societies rigid and unable to pivot as local contexts change. Government overreach would be rampant. The policies pushed from this context, detached from the on-the-ground realities, would mostly be ineffective and counterproductive. Centralized governments would mainly be self-serving and tend to reproduce the same root causes—overshoot, diminishing returns, and the paradigm of separation—and ultimately follow the same trajectory of self-destruction. The main difference would be additional suffering and damage, and an earlier unraveling.

Some people believe that preserving the natural world justifies greatly restricting our individual and collective freedoms, but this completely contradicts the core principles of life. There's a reason nature gives relative freedom to all living beings. Authoritarianism fails to recognize that domination, managerialism, and coercion are part of the root causes of the polycrisis. Without freedom, we're digging deeper into the hole. Freedom is life's way. It's the way of creation. Control, coercion, domination, and centralization are paths to extinction.

Authoritarianism and government overreach are already generating suspicion of and backlash against environmental initiatives—and understandably so. Coercive enforcement of centralized measures (that are not well thought out or sincerely intentioned) will invite a corresponding massive resistance to them. Centralization plants the seeds of its own demise.

The human path aligned with life is emergent and organic. It's not a means to an end. The path is an end in itself. This is the way forward. Life and nature are organized in an emergent way; for this reason, they are sustainable. Centralization always becomes self-serving, and in the reality of interbeing, operating mostly from a mindset of separation will always lead to destruction.

Accelerationism

Considering these predicaments, it may seem reasonable to ramp up efforts to tackle our many challenges rapidly. But this accelerationism is narrow-minded as well. Our collective trajectory, its unintended consequences, and our narrow-mindedness got us here in the first place.

The more we are challenged with complex issues, the more we need to wise up. Many people may sense trouble approaching and enter a fight-or-flight mode. However, the reckless acceleration of efforts that do not directly address the root causes of the crises will only cause more trouble. Most efforts supposedly counteracting the crises are merely a masking of symptoms, a distraction, or more of the same with different packaging—or they could become a crisis themselves. We should slow down, stay calm, and embody deep wisdom, not gamble with our future.

12. A Perfect Storm

"Everywhere people ask: 'What can I actually do?' The answer is as simple as it is disconcerting: we can, each of us, work to put our own inner house in order. The guidance we need for this work cannot be found in science or technology, the value of which utterly depends on the ends they serve; but it can still be found in the traditional wisdom of mankind." E. F. Schumacher

The elements of a perfect storm are gathering and strengthening each other. This polycrisis will eventually bring civilization to its knees. There will be a significant simplification of our modern world. It's not a matter of if but when—and this statement is not arrogance or based on fringe assumptions; it's based on the universal dynamics of complex systems. All civilizations collapse, and evidence of this lies in their countless ruins.

The question of timing is quite relevant for the current generations. A great reckoning will likely happen this century. That doesn't necessarily imply an abrupt collapse; it could be a long, staircase-style catabolic decline. The most realistic and likely scenario would be a gradual staircase decline involving sudden simplifications. The main timing indicators are the diminishing returns on complexity and resources, global warming, and other ecological impacts.

Global civilization has become very resilient and robust. A single crisis won't cause its fall. It will die from a thousand cuts, and these wounds will be internal and external. The cost of this civilization's increased sturdiness (complexity) is a trade-off: it encourages a relatively rapid collapse compared to past civilizations. However, that trade-off could be partially counteracted by the vast resilience granted by its global scale.

The groundwork of this unraveling is already being laid out, and these crises will continue to grow and erode the vitality of modernity. The internal conditions are diminishing returns on organizational and technological complexity, socioeconomic imbalances, lack of self-moderation, overshoot, war, and the paradigm of separation. The external conditions are energy and resource constraints, global warming, and the health of the living world. Most of these challenges act in self-reinforcing feedback loops, exacerbating the polycrisis. When these issues are considered holistically, it's clear that civilization will be tested to its limits. This century, humanity is in the process of deciding the amount of warming and damage to the living world we will inflict as civilization declines. The impacts we are being committed to will echo for centuries.

What happens when a *seemingly* unstoppable force meets an immovable object? When exponential complexity encounters diminishing returns and natural limits? Collapse is the most likely scenario. If humanity was working together effectively on this and not in a deadlock, perhaps it wouldn't be too farfetched to believe that we could stabilize society or engineer a more gradual and planned descent. Even if we could do that, due to the nature of this complex system, bracing for impact would likely trigger a simplification on its own.

The possibility of collapse is making its way into mainstream consciousness. In recent decades, it's been obvious how disjointed and ineffective our global efforts have been when addressing issues such as wars, pandemics, climate change, and so on. Nothing suggests that, as the core issues of the polycrisis worsen and grow synergistically, humanity will become more effective at handling them.

Yet governments can't bring themselves to digest the severity of our polycrisis and talk openly about it. Private interests spend millions to cast shadows over the information we already wish wasn't true. We know we must act, but we don't yet feel the urgency in our bones.

We must acknowledge that modernity is harming future generations of humans and other-than-humans. It has put us on a path that is too rigid and allows us little leeway or say. The modern experiment has degenerated into something contrary to the sacred will of life. Facing collapse may throw us into a crisis of meaning, causing us to question who we are, why we are here, the purpose of it all, and our relationship to life and everything around us. This is a powerful opportunity to reexamine our lives.

The most reasonable path forward would be to openly consider the full ramifications of our situation and choose to stop making things worse. It wouldn't be easy, but it would be the wisest move. It's too late to avoid massive disruptions to species and habitats, but this awareness is no excuse for nihilism or selfishness. There's lots to do, and ends are new beginnings.

For many people, realizing that we should expect a century of tribulations is not an inspiring and motivating thought. Many believe that we need optimistic, feel-good stories. But I've seen many people in difficult circumstances who don't need to hold on to such stories; such people are everywhere around us. These people keep going because they have no alternative.

It's time to take a hard look at our predicament and let go of the childish naivety of thinking that without a clear, colorful vision of everlasting happiness, we can't accomplish what needs to be done.

Not only is it intellectually dishonest if we fail to truly listen to diverse analyses of our predicament, even if we disagree with them; it also goes counter to the survival of our species. The more we look away from our challenges and keep them in the shadows, the stronger they become. Not talking about them for fear of losing control of the narrative or causing panic is the kind of response that can only worsen things. We shouldn't bury our heads in the sand when we hear people saying inconvenient things that make us uncomfortable or with which we disagree.

Younger generations are noticing that they will have to struggle with the impacts of the polycrisis, and they are often receptive to realistic and honest assessments of our situation.

Beyond Passive Hope

Hope, like motivation, can be fleeting. Hope is not a solid foundation on which to base our actions in today's world. More than hope, we need love for the miracle of life. Love entails courage, sacrifice, faithfulness, honesty, struggle, responsibility, and giving back.

The flaw in hope is its passivity. Its traditional meaning is basically about wishing and waiting for specific outcomes. The genuinely helpful concept in our context is "active hope," a term coined by Joanna Macy.[143] Active hope is about recognizing what's important to us and taking the steps to make it a reality. There's no passivity in it and no need for optimism. Active hope is about doing, not about something we identify with. To avoid being trapped in despair, active hope can serve as a means to express the desires that come from the depths of our hearts through action.

We may gain an invigorating sense of agency, responsibility, courage, and freedom when we go beyond hope. As Charles Eisenstein writes, "It is as if giving up on saving the world opens us up to doing the things that will save the world."[144] The catharsis of right action might be the stage missing from our journeys with grief.

Even if we were in a hopeless situation, that wouldn't take away the meaning of life. It wouldn't prevent us from living with integrity. The need for certainty makes many of us gravitate towards the extremes of blind despair or hope. We either think that everything is done for and that nothing we do will make a difference, or we believe everything will work out without the need for immense sacrifice. These views lack a strong impulse to act either way.

Those who cling to hope are clinging to their ideas about how the world works and how things should unfold. They fail to recognize that these projections and this mental grasping obstruct their view of the severity of our predicament and the true complexity of the world.

The idea of progress—that the human condition is improving—is one of the building blocks of modernity. Thus, many believe technology, a full-scale mobilization, or a paradigm shift will save modernity. But these savior myths are mental traps. They are spells keeping our minds—driven by our desires and fears—focused on misguided goals. In this way, our hopes fuel our destruction, and our fears serve as a springboard for those hopes. Both keep us feeding misguided agendas without noticing their destructive nature. Our hopes and fears can gradually lead us into the mindset of "we know best, we are right, we are superior." That mindset is a slippery slope into the abyss.

Clinging to naive narratives of success and minimal hardship is as much a denial as denying that we are living amidst multiple converging crises. It's not about giving up and curling ourselves into the fetal position either. Being honest with ourselves about the severity of our metacrisis doesn't entail giving up. We should do what we believe is right and what is needed of us—the doing and the journey are the reward.

We don't need hope to face the present with courage. If we're hanging on to the hope that there won't be catastrophes, we're hanging on to a delusion. Many disasters can't be prevented; our course of action is reduced to trying to survive them with as much wisdom and integrity as possible.

Embodying integrity means facing the uncertainty of reality not with a feeble hope for an optimistic outcome, but by expressing care and love without expecting a specific result. Our power doesn't lie in managing each other's perceptions through public relations and carefully crafted narratives. Our individual and collective power lies in our love, gratitude, and care for the people around us, the living beings around us, and the living world.

Hope is fragile. This is evident for those who are wholeheartedly engaged. Once we see that hope is hollow, when we release our fear and hope and let ourselves fall, we may find something much more powerful underneath to hold us. We should ground ourselves in life and interbeing as we undergo this plunge. Ultimately, what will provide us with a deeper source of sustenance will be unique to each of us, but there's no doubt that it will be stronger than optimism, hope, or fear.

Life is a gift, but many of us fail to continually recognize this. Being out of touch with nature and our mortality has allowed humanity to drift into this dark night. Doing what is right, regardless of our fears and discomfort, can be incredibly cathartic and empowering. It may trigger a personal life-changing awakening from our sleepwalking. *Right action* can help us realize we are responsible for our own fate. Not only is it our right to co-create our present and future, but it's our responsibility to honor that role. No matter how bad things could get, we can always do something to make things better. For example, we can slow down or stop the damage to the living world, people, and communities. We can better understand our situation and how to respond, and we can seek inner change at both the individual and collective levels.[145]

Emergence

Modernity trains us to think in linear cause-and-effect terms. This is useful for manipulating things on a small scale. But that linear framework is not how things work in massively complex systems like the one we take part in.

To navigate these storms, we must be flexible and adaptable. That's why allowing changes to emerge from the bottom up is so important. Efforts from the top to control and manage will only result in rigidity and misguidance. Their far distance from on-the-ground issues prevents any one-sided, top-down leadership from responding effectively in this rapidly shifting and complex context.

The wise move is to humbly accept this "unknowing." The guiding principles suggested for embracing this complexity are love, truth, freedom, interbeing, aliveness, and letting go of outcome-seeking and control. Those insights will help anyone embody authentic leadership.

We're told we need inspiring visions of the future, bold collective plans, and the will to implement them. But things are much more emergent and organic in a complex, unpredictable system. There's no need to have precise top-down plans for humanity. Those who believe we need centrally enforced plans are projecting their shadows of perceived impotence and fear of the unknown. As Mike Tyson said, everyone has a plan until they are punched in the face.

It goes against the modern paradigm of domination and control, but the best we can do is align ourselves with reality and let it unfold according to its will. It may be hard to accept this, but that's truly the best we can do. To better align with reality, we must declutter our individual and collective mindsets to clear out the misguided ideas, beliefs, and feelings that influence our lives. These forces are constricting our perception, our thoughts, and our actions. How do we align ourselves with reality? We must shift our perspective from focusing solely on how to respond for our benefit or for our survival as individuals and as a collective, to perceiving reality—what is happening, what is likely to happen, and what we can actually do about it.

We should stop spinning the issues according to our ideas and motives; we simply need to face what is. That's all. There's no need to spin anything. The best we can do is act in accordance with the will of life and reality—or, dare I say, the sacred. That's the mature and wise response. It's not just about us. It's about the miracles of existence, life, and the human species. What will we do with these miracles?

It's not about collectively sitting down and letting the status quo be; it's about processing reality and acting according to our own natures and the ancestral insights that have stood the test of time. We won't have a single unanimous voice, and that's all right. It's up to us to be the hands and voice of life. The divine is in all of us, and we are creators. It's up to us to co-create our emergent future. This breakdown may be an opportunity for massive change and the creation of alternative ways of relating and living.

What Now?

Many people will go through a rude awakening involving moments of despair, apathy, and depression, but these moments shall pass too. Afterwards, they can make space for a connection to a deeper foundation for thoughts and actions. As civilization declines many things will worsen, but many things will also improve.

Delaying and repressing a rude awakening in ourselves and others keeps us sleepwalking towards the abyss. The sleepwalkers seem to prefer the dream state, and at a deep level, they may fear realizing that they have been living in a dream. But we are all in this together, and fears and egos are no match for truth and love. The discourse of hope and fear is ultimately irrelevant. What matters are our inner and outer responses.

Regardless of our responses, there will probably be a staircase-style breakdown of civilization. It's to be expected from a complex system that has undergone unsustainable growth in size and complexity. This acceleration of the system took place without self-restraint, without considering its precarious foundation on limited resources, and without understanding its relationships with what sustains it. Understanding our situation is important, but awareness won't be very beneficial without adequate responses. The things we do individually and collectively will determine the timing and severity of this unfolding.

If we're confused about what we can do, we can ask ourselves: What do I love? And what am I going to do to carry it into the future? What good can I still do?

We must decide whether to continue propping up modernity by sacrificing the living world and our well-being—and, in the long run, modern civilization—or to take the alternative path of hospicing modernity and preventing the sacrifice of *most* of what we hold dear. We are experiencing this unraveling because our modern views, industrial civilization, and all their baggage are undermining the foundations of the living world and humanity.

Modernity has created a barrier between people and the foundations of life, partially hiding the toxic implications of our modern lives and their crumbling foundations. We are throwing future generations of humans and other-than-humans under the bus while we distract ourselves with bread and circuses.

Continuing down this path, modern civilization will persist in degrading the living world and our humanity until modernity crumbles. The likely fatal wounds could stem from the increasing unviability of large-scale agriculture combined with systemic breakdown, widespread wars, and pandemics. These wounds could be the checkmate that disrupts the functioning of modernity and civilization. This path is the business-as-usual, "managed" collapse nightmare.

The alternative path leads to fewer hardships: We let wrong views and misguided systems decay. These views and systems are in decline, but they are increasingly being propped up. The alternative path minimizes the damage to the natural world and the suffering and violence baked in and increases the quality of life for ourselves, our children, and their children. This is the path of mutual aid and of hospicing modernity. The more we embrace these attitudes individually and collectively, the better our catastrophes will be.

Many Indigenous Peoples understand that, as humans, we need a healthy foundation based on a sacred connection to the land. They know that genuine sustainability entails living in accordance with values rooted in natural laws and teachings that acknowledge the sacredness of nature; these ways of life involve sacred duties and responsibilities.[146] We must be guided by nature if we wish to survive as a species, but we won't be able to connect with its intelligent guidance if we refuse to meet Mother Earth on her terms.

We don't need the certainty that everything will improve, but the courage to get moving and forge new possibilities.[147] We must remain balanced between the raw uncertainty of the future and the conviction to create the world our hearts desire. Being content with stable societies with decent values is wiser than clinging to the boom and bust of civilizations based on a paradigm of separation without regard for life. The answer to our problem is to see who has the problem. These forces will run their course, and we don't have to bear all their weight on our individual shoulders.

There is meaning wherever we look. As our societies decline, we're offered the opportunity to rethink how and why humanity will continue. The slowing down of the mega-machine and its systematic destruction of life is not that far off in the future. This era of loss is also an era of rebirth. We need wisdom and compassion to overcome these challenges without tearing ourselves apart. Nothing in our future is entirely determined. We're all entangled in this unfolding, so the future is ours for the making.

Notes

1   Andreotti, Vanessa. 2023. "Unlearning Bundle (Unit) 1: What is Education for?" Facing Human Wrongs. https://facinghumanwrongs.net/unlearning-bundle-unit-1/.

2   Chefurka, Paul. 2012. "Climbing The Ladder of Awareness." Paul Chefurka. http://www.paulchefurka.ca/LadderOfAwareness.html/.

3   Machado de Oliveira, Vanessa. 2021. *Hospicing Modernity: Facing Humanity's Wrongs and the Implications for Social Activism*. Berkeley: North Atlantic Books, 47.

4   *The Red Book: A Reader's Edition*. This more or less refers to the ego attachments of our hero, and our projections of God, not the sacred.

5   Becker, Ernest. 1997. *The Denial of Death*. New York: Free Press.

6   van Kessel, Cathryn, Nicholas Jacobs, Francesca Catena, and Kimberly Edmondson. "Responding to Worldview Threats in the Classroom: An Exploratory Study of Preservice Teachers." *Journal of Teacher Education* 73, no. 1 (2022): 97–109. https://doi.org/10.1177/00224871211051991/.

7   Boyd, Andrew. 2023. *I Want a Better Catastrophe: Navigating the Climate Crisis with Grief, Hope, and Gallows Humor*. Gabriola Island: New Society Publishers, 68.

8   Machado de Oliveira, *Hospicing Modernity*, 23.

9   Servigne, Pablo, Raphaël Stevens, and Gauthier Chapelle. 2021. *Another End of the World Is Possible: Living the Collapse (and Not Merely Surviving It)*. Translated by Geoffrey Samuel. N.p.: Polity Press, 58.

10   Servigne, Stevens, and Chapelle, *Another End of the World Is Possible*, 58.

11   Ross, Rick. 2009. "Watch out for tell-tale signs." *The Guardian*, May 27, 2009. http://theguardian.com/commentisfree/belief/2009/may/27/cults-definition-religion/.

12   Kramer, Joel, and Diana Alstad. 1993. *The Guru Papers: Masks of Authoritarian Power*. N.p.: North Atlantic Books.

13   Boyd, *I Want a Better Catastrophe*, 205.

14   "At the Killing Centers." 2023. Holocaust Encyclopedia. http://encyclopedia.ushmm.org/content/en/article/at-the-killing-centers/.

15   Bendell, Jem. 2018. "Barriers to Dialogue on Deep Adaptation." Prof Jem Bendell. https://jembendell.com/2018/08/20/barriers-to-dialogue-on-deep-adaptation/.

16   Diamond, Jared. 2011. *Collapse: How Societies Choose to Fail Or Succeed: Revised Edition*. New York: Penguin Publishing Group.

17   Sundstrom, Shana M., and Craig R. Allen. "The Adaptive Cycle: More Than a Metaphor." *Ecological Complexity* 39 (2019): 100767. https://www.sciencedirect.com/science/article/pii/S1476945X1830165X/.

18   Cumming, Graeme S., and Garry D. Peterson. "Unifying Research on Social–Ecological Resilience and Collapse." *Trends in Ecology & Evolution* 32, no. 9 (2017): 695–713. https://mahb.stanford.edu/wp-content/uploads/2019/01/cumming2017_UnifyingResearchonSocial-EcologicalResilience.pdf/.

19   Servigne, Stevens, and Chapelle, *Another End of the World Is Possible*, 83.

20   Burkhard, Benjamin, Brian D. Fath, and Felix Müller. "Adapting the Adaptive Cycle: Hypotheses on the Development of Ecosystem Properties and Services." *Ecological Modelling* 222, no. 16 (2011): 2878–2890. https://www.sciencedirect.com/science/article/abs/pii/S0304380011002961/.

21   Boyd, *I Want a Better Catastrophe*, 368.

22   Bardi, Ugo, Sara Falsini, and Ilaria Perissi. "Toward a General Theory of Societal Collapse: A Biophysical Examination of Tainter's Model of the Diminishing Returns of Complexity." *BioPhysical Economics and Resource Quality* 4 (2019): 1–9. https://link.springer.com/article/10.1007/s41247-018-0049-0/.

23   Scheffer, Marten, Jordi Bascompte, William A. Brock, Victor Brovkin, Stephen R. Carpenter, Vasilis Dakos, Hermann Held, Egbert H. Van Nes, Max Rietkerk, and George Sugihara. "Early-Warning Signals for Critical Transitions." *Nature* 461, no. 7260 (2009): 53–59. https://pdodds.w3.uvm.edu/files/papers/others/2009/scheffer2009a.pdf/.

24   Dakos, Vasilis, Egbert H. van Nes, and Marten Scheffer. "Flickering as an Early Warning Signal." *Theoretical Ecology* 6 (2013): 309–317. https://link.springer.com/article/10.1007/s12080-013-0186-4/.

25   Boyd, *I Want a Better Catastrophe*, 55.

26   Feng, Shuo, Chen Shen, Nan Xia, Wei Song, Mengzhen Fan, and Benjamin J. Cowling. "Rational Use of Face Masks in the COVID-19 Pandemic." *The Lancet Respiratory Medicine* 8, no. 5 (2020): 434–436. https://www.thelancet.

com/journals/lanres/article/PIIS2213-2600(20)30134-X/fulltext/; Dyer, Evan. 2020. "Some health experts questioning advice against wider use of masks to slow spread of COVID-19." *CBC*, April 28, 2020. https://www.cbc.ca/news/politics/covid-19-pandemic-coronavirus-masks-1.5515526/.

27 Hart, Justin. 2022. "The Twitter Blacklisting of Jay Bhattacharya." *The Wall Street Journal*, December 9, 2022. https://www.wsj.com/articles/the-twitter-blacklisting-of-jay-bhattacharya-medical-expert-covid-lockdown-stanford-doctor-shadow-banned-censorship-11670621083/.

28 Khurshudyan, Isabelle. 2022. "An interview with Ukrainian President Volodymyr Zelensky." *The Washington Post*, August 23, 2022. https://www.washingtonpost.com/national-security/2022/08/16/zelensky-interview-transcript/.

29 McGilchrist, Iain. 2012. *The Master and His Emissary: The Divided Brain and the Making of the Western World*. N.p.: Yale University Press.

30 RSA. 2011. "RSA ANIMATE: The Divided Brain." YouTube. https://www.youtube.com/watch?v=dFs9WO2B8uI/.

31 Poore, Joseph, and Thomas Nemecek. "Reducing Food's Environmental Impacts Through Producers and Consumers." *Science* 360, no. 6392 (2018): 987–992. https://www.science.org/doi/10.1126/science.aaq0216/.

32 Folke, Carl, Stephen Polasky, Johan Rockström, Victor Galaz, Frances Westley, Michèle Lamont, Marten Scheffer et al. "Our Future in the Anthropocene Biosphere." *Ambio* 50 (2021): 834–869. https://link.springer.com/article/10.1007/s13280-021-01544-8/.

33 "Norman Borlaug Nobel Lecture." 1970. The Nobel Prize. nobelprize.org/prizes/peace/1970/borlaug/lecture/.

34 "UN projects world population to peak within this century." 2024. United Nations. http://un.org/en/UN-projects-world-population-to-peak-within-this-century/.

35 Catton, William R. 1982. *Overshoot: The Ecological Basis of Revolutionary Change*. N.p.: University of Illinois Press.

36 Catton, *Overshoot*.

37 Scheffer, Victor B. "The Rise and Fall of a Reindeer Herd." *The Scientific Monthly* 73, no. 6 (1951): 356–362. https://www.jstor.org/stable/20582/.

38 Ritchie, Hannah. 2024. "One-third of the world's assessed fish stocks are overexploited." Our World in Data. https://ourworldindata.org/data-insights/one-third-of-the-worlds-assessed-fish-stocks-are-overexploited/.

39 Wagner, David L., Eliza M. Grames, Matthew L. Forister, May R. Berenbaum, and David Stopak. "Insect Decline in the Anthropocene: Death by a Thousand Cuts." *Proceedings of the National Academy of Sciences* 118, no. 2 (2021): e2023989118. https://www.pnas.org/doi/full/10.1073/pnas.2023989118/.

40 Curtis, Philip G., Christy M. Slay, Nancy L. Harris, Alexandra Tyukavina, and Matthew C. Hansen. "Classifying Drivers of Global Forest Loss." *Science* 361, no. 6407 (2018): 1108–1111. https://www.science.org/doi/10.1126/science.aau3445/.

41 Fanning, Andrew L., Daniel W. O'Neill, Jason Hickel, and Nicolas Roux. "The Social Shortfall and Ecological Overshoot of Nations." *Nature Sustainability* 5, no. 1 (2022): 26–36. https://www.nature.com/articles/s41893-021-00799-z/.

42 Tainter, Joseph. 1988. *The Collapse of Complex Societies*. N.p.: Cambridge University Press.

43 Bardi, Falsini, and Perissi, "Toward a General Theory of Societal Collapse."

44 Bardi, Falsini, and Perissi, "Toward a General Theory of Societal Collapse."

45 Jancovici, Jean-Marc. 2013. "How much of a slave master am I?" Jean-Marc Jancovici. https://jancovici.com/en/energy-transition/energy-and-us/how-much-of-a-slave-master-am-i/.

46 Ritchie, Hannah, and Pablo Rosado. 2024. "Energy Mix." Our World in Data. https://ourworldindata.org/energy-mix/.

47 Berman, Art. 2023. "They're Not Making Oil Like They Used To: Stealth Peak Oil?" Art Berman. artberman.com/blog/theyre-not-making-oil-like-they-used-to-stealth-peak-oil/.

48 Geological Survey of Finland. 2022. "GTK Research: The Currently Known Global Mineral Reserves Will Not Be Sufficient to Supply Enough Metals to Manufacture the Planned Non-fossil Fuel Industrial Systems." GTK. https://www.gtk.fi/en/current/gtk-research-the-currently-known-global-mineral-reserves-will-not-be-sufficient-to-supply-enough-metals-to-manufacture-the-planned-non-fossil-fuel-industrial-systems/.

49 Berman, "They're Not Making Oil Like They Used To."

50 Energy Institute. 2024. "Oil Production." Our World in Data. ourworldindata.org/grapher/oil-production-by-country?time=1926..latest&country=~OWID_WRL/.

51 Hall, Charles A.S., Jessica G. Lambert, and Stephen B. Balogh. "EROI of Different Fuels and the Implications for Society." *Energy Policy* 64 (2014): 141–152. https://doi.org/10.1016/j.enpol.2013.05.049/.

52 Court, Victor, and Florian Fizaine. "Long-Term Estimates of the Energy-Return-on-Investment (EROI) of Coal, Oil, and Gas Global Productions." *Ecological Economics* 138 (2017): 145–159. https://doi.org/10.1016/j.ecolecon.2017.03.015/.

53 Heun, Matthew Kuperus, and Martin de Wit. "Energy Return on (Energy) Invested (EROI), Oil Prices, and Energy Transitions." *Energy Policy* 40 (2012): 147–158. https://doi.org/10.1016/j.enpol.2011.09.008/.

54 Jackson, Andrew, and Tim Jackson. "Modelling Energy Transition Risk: The Impact of Declining Energy Return on Investment (EROI)." *Ecological Economics*

185 (2021): 107023. https://doi.org/10.1016/j.ecolecon.2021.107023/; Brockway, Paul E., Anne Owen, Lina I. Brand-Correa, and Lukas Hardt. "Estimation of Global Final-Stage Energy-Return-on-Investment for Fossil Fuels With Comparison to Renewable Energy Sources." *Nature Energy* 4, no. 7 (2019): 612–621. https://doi.org/10.1038/s41560-019-0425-z/.

55   Geological Survey of Finland. 2022. "There Are Bottlenecks in Raw Materials Supply Chain – A Glimpse of the Systemic Overview Is Here, Discussion and the Development of the Solutions Have Started." GTK. https://www.gtk.fi/en/current/there-are-bottlenecks-in-raw-materials-supply-chain-a-glimpse-of-the-systemic-overview-is-here-discussion-and-the-development-of-the-solutions-have-started/.

56   IEA. 2024. "IEA." Executive summary – Electricity 2024. https://www.iea.org/reports/electricity-2024/executive-summary/.

57   Turchin, Peter. 2024. *End Times: Elites, Counter-Elites and the Path of Political Disintegration*. N.p.: Penguin Publishing Group.

58   Oxfam. 2022. "Ten richest men double their fortunes in pandemic while incomes of 99 percent of humanity fall." Oxfam. oxfam.org/en/press-releases/ten-richest-men-double-their-fortunes-pandemic-while-incomes-99-percent-humanity/.

59   Turchin, *End Times*.

60   FAO. 2024. "Calorie supply by food group, World, 2022." Our World in Data. ourworldindata.org/grapher/calorie-supply-by-food-group?country=~OWID_WRL/.

61   Our World in Data. 2015. "World population with and without synthetic nitrogen fertilizers." Our World in Data. https://ourworldindata.org/grapher/world-population-with-and-without-fertilizer/.

62   IEA. 2021. "Ammonia Technology Roadmap Executive Summary." IEA. https://www.iea.org/reports/ammonia-technology-roadmap/executive-summary/.

63   Samreen, Sayma, and Sharba Kausar. "Phosphorus Fertilizer: The Original and Commercial." *Phosphorus: Recovery and Recycling* (2019): 81. https://www.intechopen.com/chapters/64614/.

64   Torrella, Kenny. 2022. "Sri Lanka's organic farming disaster, explained." *Vox*, July 15, 2022. http://vox.com/future-perfect/2022/7/15/23218969/sri-lanka-organic-fertilizer-pesticide-agriculture-farming/.

65   Zhao, Chuang, Bing Liu, Shilong Piao, Xuhui Wang, David B. Lobell, Yao Huang, Mengtian Huang et al. "Temperature Increase Reduces Global Yields of Major Crops in Four Independent Estimates." *Proceedings of the National Academy of sciences* 114, no. 35 (2017): 9326–9331. https://doi.org/10.1073/pnas.1701762114/.

66   Iizumi, Toshichika, Jun Furuya, Zhihong Shen, Wonsik Kim, Masashi Okada,

Shinichiro Fujimori, Tomoko Hasegawa, and Motoki Nishimori. "Responses of Crop Yield Growth to Global Temperature and Socioeconomic Changes." *Scientific Reports* 7, no. 1 (2017): 7800. https://www.nature.com/articles/s41598-017-08214-4/; Levis, Samuel, Andrew Badger, Beth Drewniak, Cynthia Nevison, and Xiaolin Ren. "CLMcrop Yields and Water Requirements: Avoided Impacts by Choosing RCP 4.5 over 8.5." *Climatic Change* 146 (2018): 501–515. https://doi.org/10.1007/s10584-016-1654-9/.

67   Iizumi, Furuya, Shen, Kim, Okada, Fujimori, Hasegawa, and Nishimori, "Responses of Crop Yield Growth to Global Temperature and Socioeconomic Changes."

68   Iizumi, Furuya, Shen, Kim, Okada, Fujimori, Hasegawa, and Nishimori, "Responses of Crop Yield Growth to Global Temperature and Socioeconomic Changes."

69   Fuglie, Keith. 2023. "Growth rate of global agricultural output has slowed." Economic Research Service. https://www.ers.usda.gov/data-products/charts-of-note/chart-detail?chartId=107931/.

70   FAO. 2025. "FAOSTAT Database." Food and Agriculture Organization of the United Nations. https://www.fao.org/faostat/en/#data/SDGB/; Ritchie, Hannah, Pablo Rosado, and Max Roser. 2023. "Hunger and Undernourishment." Our World in Data. http://ourworldindata.org/hunger-and-undernourishment/.

71   Barnosky, Anthony D., Nicholas Matzke, Susumu Tomiya, Guinevere O.U. Wogan, Brian Swartz, Tiago B. Quental, Charles Marshall et al. "Has the Earth's Sixth Mass Extinction Already Arrived?" *Nature* 471, no. 7336 (2011): 51–57. https://doi.org/10.1038/nature09678/.

72   Brannen, Peter. 2017. "Earth Is Not in the Midst of a Sixth Mass Extinction." *The Atlantic*, June 13, 2017. http://theatlantic.com/science/archive/2017/06/the-ends-of-the-world/529545/.

73   Ceballos, Gerardo, Paul R. Ehrlich, Anthony D. Barnosky, Andrés García, Robert M. Pringle, and Todd M. Palmer. "Accelerated Modern Human-Induced Species Losses: Entering the Sixth Mass Extinction." *Science Advances* 1, no. 5 (2015): e1400253. https://doi.org/10.1126/sciadv.1400253/.

74   Greenspoon, Lior, Eyal Krieger, Ron Sender, Yuval Rosenberg, Yinon M. Bar-On, Uri Moran, Tomer Antman et al. "The Global Biomass of Wild Mammals." *Proceedings of the National Academy of Sciences* 120, no. 10 (2023): e2204892120. https://doi.org/10.1073/pnas.2204892120/.

75   IPBES. 2019. "Nature's Dangerous Decline 'Unprecedented'; Species Extinction Rates 'Accelerating.'" IPBES. https://www.ipbes.net/news/Media-Release-Global-Assessment/.

76   Newbold, Tim. "Future Effects of Climate and Land-Use Change on Terrestrial Vertebrate Community Diversity Under Different Scenarios." *Proceedings*

of the Royal Society B 285, no. 1881 (2018): 20180792. https://doi.org/10.1098/rspb.2018.0792/.

77   Richardson, Katherine, Will Steffen, Wolfgang Lucht, Jørgen Bendtsen, Sarah E. Cornell, Jonathan F. Donges, Markus Drüke et al. "Earth Beyond Six of Nine Planetary Boundaries." *Science Advances* 9, no. 37 (2023): eadh2458. https://doi.org/10.1126/sciadv.adh2458/.

78   Caesar, Levke, Boris Sakschewski, Lauren Seaby Andersen, Tim Beringer, Johanna Braun, Donovan Dennis, Dieter Gerten et al. 2024. "Planetary Health Check 2024: A Scientific Assessment of the State of the Planet." https://www.planetaryhealthcheck.org/storyblok-cdn/f/301438/x/a4efc3f6d5/planetary-healthcheck2024_report.pdf/.

79   Richardson, Steffen, Lucht, Bendtsen, Cornell, Donges, Drüke et al., "Earth Beyond Six of Nine Planetary Boundaries."

80   EPA. 2025. "Climate Change Indicators: Ocean Acidity." Environmental Protection Agency. http://epa.gov/climate-indicators/climate-change-indicators-ocean-acidity/.

81   Stockholm Resilience Centre. 2025. "Planetary boundaries." Stockholm Resilience Centre. https://www.stockholmresilience.org/research/planetary-boundaries.html/.

82   Hansen, James E., Makiko Sato, Leon Simons, Larissa S. Nazarenko, Isabelle Sangha, Pushker Kharecha, James C. Zachos et al. "Global Warming in the Pipeline." *Oxford Open Climate Change* 3, no. 1 (2023): kgad008. https://doi.org/10.1093/oxfclm/kgad008/.

83   Rickaby, Rosalind. 2022. "When hippos roamed, past temperatures were much higher." *University of Oxford*, July 19, 2022. https://www.ox.ac.uk/news/2022-07-19-when-hippos-roamed-past-temperatures-were-much-higher-expert-comment/.

84   Hansen, James E., Pushker Kharecha, Makiko Sato, George Tselioudis, Joseph Kelly, Susanne E. Bauer, Reto Ruedy et al. "Global Warming Has Accelerated: Are the United Nations and the Public Well-Informed?" *Environment: Science and Policy for Sustainable Development* 67, no. 1 (2025): 6–44. https://doi.org/10.1080/00139157.2025.2434494/.

85   Kulp, Scott A., and Benjamin H. Strauss. "New Elevation Data Triple Estimates of Global Vulnerability to Sea-Level Rise and Coastal Flooding." *Nature Communications* 10, no. 1 (2019): 1–12. https://doi.org/10.1038/s41467-019-12808-z/.

86   Carlowicz, Michael. 2022. "Tracking 30 Years of Sea Level Rise." NASA Earth Observatory. http://earthobservatory.nasa.gov/images/150192/tracking-30-years-of-sea-level-rise/.

87   NOAA. 2025. "Trends in CO2." NOAA Global Monitoring Laboratory. https://gml.noaa.gov/ccgg/trends/.

88 UNCCD. 2022. "Land-based solutions offer a key opportunity for climate mitigation." UNCCD. unccd.int/news-stories/stories/land-based-solutions-offer-key-opportunity-climate-mitigation/.

89 Keeling, Ralph F., and Heather D. Graven. "Insights From Time Series of Atmospheric Carbon Dioxide and Related Tracers." *Annual Review of Environment and Resources* 46, no. 1 (2021): 85–110. https://doi.org/10.1146/annurev-environ-012220-125406/.

90 Curran, James C., and Samuel A. Curran. "Natural Sequestration of Carbon Dioxide Is in Decline: Climate Change Will Accelerate." *Weather* 80, no. 3 (2025): 85–87. https://doi.org/10.1002/wea.7668/.

91 Curran, James C., and Samuel A. Curran. "Indications of Positive Feedback in Climate Change Due to a Reduction in Northern Hemisphere Biomass Uptake of Atmospheric Carbon Dioxide." *Weather* 71, no. 4 (2016): 88–91. https://doi.org/10.1002/wea.2715/.

92 Wang, Seaver, Adrianna Foster, Elizabeth A. Lenz, John D. Kessler, Julienne C. Stroeve, Liana O. Anderson, Merritt Turetsky et al. "Mechanisms and Impacts of Earth System Tipping Elements." *Reviews of Geophysics* 61, no. 1 (2023): e2021RG000757. https://doi.org/10.1029/2021RG000757/; Wunderling, Nico, Anna S. von der Heydt, Yevgeny Aksenov, Stephen Barker, Robbin Bastiaansen, Victor Brovkin, Maura Brunetti et al. "Climate Tipping Point Interactions and Cascades: A Review." *Earth System Dynamics* 15, no. 1 (2024): 41–74. https://doi.org/10.5194/esd-15-41-2024/.

93 Armstrong McKay, David I., Arie Staal, Jesse F. Abrams, Ricarda Winkelmann, Boris Sakschewski, Sina Loriani, Ingo Fetzer, Sarah E. Cornell, Johan Rockström, and Timothy M. Lenton. "Exceeding 1.5 C Global Warming Could Trigger Multiple Climate Tipping Points." *Science* 377, no. 6611 (2022): eabn7950. https://doi.org/10.1126/science.abn7950/.

94 Deutloff, Jakob, Hermann Held, and Timothy M. Lenton. "High Probability of Triggering Climate Tipping Points Under Current Policies Modestly Amplified by Amazon Dieback and Permafrost Thaw." *Earth System Dynamics* 16, no. 2 (2025): 565–583. https://doi.org/10.5194/esd-16-565-2025/.

95 Hansen, Kharecha, Sato, Tselioudis, Kelly, Bauer, Ruedy et al., "Global Warming Has Accelerated: Are the United Nations and the Public Well-Informed?"

96 Ditlevsen, Peter, and Susanne Ditlevsen. "Warning of a Forthcoming Collapse of the Atlantic Meridional Overturning Circulation." *Nature Communications* 14, no. 1 (2023): 1–12. https://www.nature.com/articles/s41467-023-39810-w/; Smolders, Emma J.V., René M. van Westen, and Henk A. Dijkstra. "Probability Estimates of a 21st Century AMOC Collapse." *arXiv preprint* arXiv:2406.11738 (2024). https://arxiv.org/html/2406.11738v1/.

97 Van Westen, René M., Michael Kliphuis, and Henk A. Dijkstra. "Physics-Based

Early Warning Signal Shows That AMOC Is on Tipping Course." *Science Advances* 10, no. 6 (2024): eadk1189. https://doi.org/10.1126/sciadv.adk1189/.

98   Armstrong McKay, Staal, Abrams, Winkelmann, Sakschewski, Loriani, Fetzer, Cornell, Rockström, and Lenton, "Exceeding 1.5 C Global Warming Could Trigger Multiple Climate Tipping Points."

99   Hansen, Kharecha, Sato, Tselioudis, Kelly, Bauer, Ruedy et al., "Global Warming Has Accelerated: Are the United Nations and the Public Well-Informed?"

100   Deutloff, Held, and Lenton, "High Probability of Triggering Climate Tipping Points Under Current Policies Modestly Amplified by Amazon Dieback and Permafrost Thaw."

101   Hansen, Kharecha, Sato, Tselioudis, Kelly, Bauer, Ruedy et al., "Global Warming Has Accelerated: Are the United Nations and the Public Well-Informed?"

102   Levy, Hiram, Larry W. Horowitz, M. Daniel Schwarzkopf, Yi Ming, Jean-Christophe Golaz, Vaishali Naik, and V. Ramaswamy. "The Roles of Aerosol Direct and Indirect Effects in Past and Future Climate Change." *Journal of Geophysical Research: Atmospheres* 118, no. 10 (2013): 4521–4532. https://doi.org/10.1002/jgrd.50192/.

103   Courchene, David, Harry Bone, Florence Paynter, Phillip Paynter, Katherine Whitecloud, Robert Maytwayashing, Mary Maytwayashing, and Gordon Walker. 2021. "Wahbanung the Resurgence of a People: Clearing the Path for our Survival: A Summary." Turtle Lodge Central House of Knowledge. https://manitobachiefs.com/wp-content/uploads/2020/07/Summary_of_WAHBANUNG_V2-Final_WEB.pdf/.

104   IPCC. 2021. "Chapter 6: Short-lived Climate Forcers." IPCC. http://ipcc.ch/report/ar6/wg1/chapter/chapter-6/.

105   NASA. 2024. "Methane – Climate Change: Vital Signs of the Planet." Climate Change NASA. http://climate.nasa.gov/vital-signs/methane/.

106   NOAA. 2023. "Greenhouse gases continued to increase rapidly in 2022." NOAA. https://www.noaa.gov/news-release/greenhouse-gases-continued-to-increase-rapidly-in-2022/.

107   Hansen, Kharecha, Sato, Tselioudis, Kelly, Bauer, Ruedy et al., "Global Warming Has Accelerated: Are the United Nations and the Public Well-Informed?"

108   Wikipedia. 2025. "Big lie." http://en.wikipedia.org/wiki/Big_lie/; Hitler, Adolf. 1992. *Mein Kampf.* Translated by Ralph Manheim. N.p.: Pimlico.

109   Hansen, Kharecha, Sato, Tselioudis, Kelly, Bauer, Ruedy et al., "Global Warming Has Accelerated: Are the United Nations and the Public Well-Informed?"

110   Larsson Blind, Åsa. 2024. "Saami Council's statement to the Harvard decision to Halt the SCoPEx Project." Saami Council. https://www.saamicouncil.net/

news-archive/saami-councils-statement-to-the-harvard-decision-to-halt-the-scopex-project/.

111   Lovelock, James. 2000. *Gaia: A New Look at Life on Earth*. N.p.: Oxford Paperbacks.

112   Thomas, Sunil, Jacques Izard, Emily Walsh, Kristen Batich, Pakawat Chongsathidkiet, Gerard Clarke, David A. Sela et al. "The Host Microbiome Regulates and Maintains Human Health: A Primer and Perspective for Non-Microbiologists." *Cancer Research* 77, no. 8 (2017): 1783–1812. https://doi.org/10.1158/0008-5472.CAN-16-2929/.

113   Ritchie, Hannah, and Max Roser. 2024. "CO2 emissions." Our World in Data. http://ourworldindata.org/co2-emissions/.

114   Parrique, Timothée, Jonathan Barth, François Briens, Christian Kerschner, Alejo Kraus-Polk, Anna Kuokkanen, and Joachim H. Spangenberg. 2019. "Decoupling Debunked: Evidence and Arguments Against Green Growth as a Sole Strategy for Sustainability." European Environment Bureau. https://eeb.org/library/decoupling-debunked/.

115   Bergmann, Tobias, and Matthias Kalkuhl. "Decoupling Economic Growth From Energy Use: The Role of Energy Intensity in an Endogenous Growth Model." *Ecological Economics* 230 (2025): 108519. https://www.sciencedirect.com/science/article/pii/S0921800925000023/; Haberl, Helmut, Dominik Wiedenhofer, Doris Virág, Gerald Kalt, Barbara Plank, Paul Brockway, Tomer Fishman et al. "A Systematic Review of the Evidence on Decoupling of GDP, Resource Use and GHG Emissions, Part II: Synthesizing the Insights." *Environmental Research Letters* 15, no. 6 (2020): 065003. https://iopscience.iop.org/article/10.1088/1748-9326/ab842a/.

116   "Norman Borlaug Nobel Lecture." 1970. The Nobel Prize. nobelprize.org/prizes/peace/1970/borlaug/lecture/.

117   Walsh, Bryan. 2023. "The doomers are wrong about humanity's future — and its past." *Vox*, March 20, 2023. https://www.vox.com/the-highlight/23627382/progress-climate-change-poverty-global-health-doom-industrial-revolution-vaccines/; Klein, Ezra. 2022. "Your Kids Are Not Doomed." *The New York Times*, June 5, 2022. https://www.nytimes.com/2022/06/05/opinion/climate-change-should-you-have-kids.html/.

118   Bertuccio, Paola, Andrea Amerio, Enrico Grande, Carlo La Vecchia, Alessandra Costanza, Andrea Aguglia, Isabella Berardelli et al. "Global Trends in Youth Suicide From 1990 to 2020: An Analysis of Data From the WHO Mortality Database." *EClinicalMedicine* 70 (2024). https://www.thelancet.com/journals/eclinm/article/PIIS2589-5370(24)00085-3/fulltext/.

119   UN FAO. 2025. "Number of people that are undernourished." Our World in

Data. https://ourworldindata.org/grapher/number-undernourished?country=~OWID_WRL/.

120 Brecke, Peter. 2023. "Global deaths in violent political conflicts over the long run." Our World in Data. https://ourworldindata.org/grapher/global-deaths-in-violent-political-conflicts-over-the-long-run?yScale=linear/.

121 Dattani, Saloni. 2023. "What were the death tolls from pandemics in history?" Our World in Data. https://ourworldindata.org/historical-pandemics/.

122 Mann, Michael. 2023. "Stop the doom. We failed to prevent climate change – but we will decide how bad it'll get." *USA Today*, September 27, 2023. https://www.usatoday.com/story/opinion/voices/2023/09/27/climate-change-hottest-summer-what-to-do/70901908007/; Theil, Michele. 2023. "Doomism about climate change is a privilege we can't afford." *Big Issue*, April 20, 2023. https://www.bigissue.com/news/environment/earth-day-doomism-the-privilege-we-can-ill-afford/.

123 Svoboda, Michael. 2023. "Renowned climate scientist Michael E. Mann on what 'doomers' get wrong." Yale Climate Connections. https://yaleclimateconnections.org/2023/09/renowned-climate-scientist-michael-e-mann-on-what-doomers-get-wrong/; Coaston, Jane. 2022. "Try to Resist the Call of the Doomers." *The New York Times*, July 23, 2022. https://www.nytimes.com/2022/07/23/opinion/climate-doomers-possibility.html/.

124 Ross, "Watch out for tell-tale signs."

125 Osaka, Shannon. 2023. "Why climate 'doomers' are replacing climate 'deniers.'" *The Washington Post*, March 24, 2023. https://www.washingtonpost.com/climate-environment/2023/03/24/climate-doomers-ipcc-un-report/.

126 Collins, Jim. n.d. "The Stockdale Paradox." Jim Collins. http://jimcollins.com/concepts/Stockdale-Concept.html/.

127 Nuttall, Philippa. n.d. "Don't listen to the climate doomists." *The New Statesman*. Accessed 2024. https://www.newstatesman.com/spotlight/sustainability/climate/2022/08/climate-doomism-dont-listen-toxic-narrative/.

128 Theil, Michele. 2023. "Doomism about climate change is a privilege we can't afford." *Big Issue*, April 20, 2023. https://www.bigissue.com/news/environment/earth-day-doomism-the-privilege-we-can-ill-afford/.

129 Servigne, Stevens, and Chapelle, *Another End of the World Is Possible*, 131.

130 Boyd, *I Want a Better Catastrophe*, 51.

131 Servigne, Stevens, and Chapelle, *Another End of the World Is Possible*, 160.

132 Manna, Antonietta, Antonino Raffone, Mauro Gianni Perrucci, Davide Nardo, Antonio Ferretti, Armando Tartaro, Alessandro Londei, Cosimo Del Gratta, Marta Olivetti Belardinelli, and Gian Luca Romani. "Neural Correlates of Focused Attention and Cognitive Monitoring in Meditation." *Brain Research*

Bulletin 82, no. 1-2 (2010): 46–56. https://www.sciencedirect.com/science/article/abs/pii/S0361923010000614/.

133   Boyd, *I Want a Better Catastrophe*, 191.

134   Servigne, Stevens, and Chapelle, *Another End of the World Is Possible*, 38.

135   Servigne, Stevens, and Chapelle, *Another End of the World Is Possible*, 54.

136   Dowd, Michael. 2023. "The Big Picture Beyond Hope and Fear." Canadian Association for the Club of Rome. https://canadiancor.com/wp-content/uploads/2023/05/Dowd-CACOR-Beyond-Hope-and-Fear-v4-1.pdf/.

137   Dattani, Saloni, Lucas Rodés, and Max Roser. 2025. "Fertility Rate." Our World in Data. https://ourworldindata.org/fertility-rate/; Smoak, Natalie, and Fatima Foflonker. 2025. "Fertility rate." Britannica. https://www.britannica.com/topic/fertility-rate/.

138   Servigne, Stevens, and Chapelle, *Another End of the World Is Possible*, 117.

139   Schrei, Joshua. 2023. "The Emerald: Oh Justice." DeepCast. deepcast.fm/episode/oh-justice/.

140   Movement Ecology Collective. 2025. "Collective Agency: Building Power in the Face of Collapse."

141   Servigne, Stevens, and Chapelle, *Another End of the World Is Possible*, 184; Bendell, Jem. 2023. *Breaking Together: A Freedom-loving Response to Collapse*. N.p.: Good Works, 428.

142   Ritchie, Hannah, Pablo Rosado, and Max Roser. 2024. "Energy Production and Consumption." Our World in Data. http://ourworldindata.org/energy-production-consumption/; Ritchie, Hannah. 2021. "Global comparison: how much energy do people consume?" Our World in Data. http://ourworldindata.org/per-capita-energy/.

143   Macy, Joanna, and Chris Johnstone. 2012. *Active Hope: How to Face the Mess We're in Without Going Crazy*. N.p.: New World Library.

144   Eisenstein, Charles. 2018. "Climate A New Story - The Climate Spectrum and Beyond." Charles Eisenstein. https://charleseisenstein.org/books/climate-a-new-story/eng/the-end-of-the-world/.

145   Servigne, Stevens, and Chapelle, *Another End of the World Is Possible*, 201.

146   Courchene, Paynter, Bone, Paynter, Whitecloud, Maytwayashing, Maytwayashing, and Walker, "Wahbanung the Resurgence of a People: Clearing the Path for our Survival: A Summary."

147   Servigne, Stevens, and Chapelle, *Another End of the World Is Possible*, 63.

Peer-Support Resources

Work That Reconnects Network (workthatreconnects.org) is a global network of facilitators (inspired by the work of Joanna Macy) that provides events, resources, and courses focused on relating to "The Great Turning" in a healing and empowering way.

Good Grief Network (goodgriefnetwork.org) is an organization that runs peer-support groups with the intention of helping people metabolize heavy emotions related to the planetary crises.

Safe Circle Calls (livingresilience.net/safecircle) are online video peer-support calls for those who would like to connect with other collapse-aware people.

Deep Adaptation Forum (deepadaptation.info) is an online mutual-support forum primarily for professional collaboration on areas related to collapse.

Zen and the Art of Saving the Planet (by Thich Nhat Hahn) is not an organization, but a book that may be quite helpful for processing and engaging with difficult thoughts and emotions and for relating to the crises in a regenerative way.

Collapse Support Reddit (reddit.com/r/collapsesupport) is an online place for discussion about collapse and about coping with related concerns. Beware, this community sometimes has a very depressing energy.

Acknowledgements

Collapse would have never been written without the encouragement, patience, and multifaceted support of my wife, Jennifer Ford. Her feedback throughout the writing process also helped improve and clarify the text of the book. I'd like to acknowledge that this book was written on hunting and gathering grounds that have been used by Indigenous Peoples for millennia. I'm very grateful for living in the boreal forest.

Thank you to the writers, speakers, and storytellers, and to the thousands of generations of humans for progressively advancing our understanding of ourselves and our world.

Thank you so much to all the alpha readers (mentioned and unmentioned) for taking the time and making the effort to not just read my unpolished drafts, but to send me detailed comments or general feedback. This book is far better thanks to you. Special thanks to Andrea Pitio, Carolyn Baker, Gail Bradbrook, Milo Gonzalez, Rob Dietz, and Ugo Bardi.

I'd like to thank Brock Peters for his attention to detail and the quality of his copy editing. I would also like to thank Christina Wulandari for creating a great cover.

I'm grateful to all those who have shared relevant ideas through recordings or writings—people such as William R. Catton, Joanna Macy, Joseph Tainter, Jared Diamond, James Lovelock, James Hansen, John Michael Greer, Alan Watts, Thich Nhat Hahn, Pablo Servigne, Raphaël Stevens, Gauthier Chapelle, William Rees, Vanessa Andreotti, Charles Eisenstein, Richard Heinberg, Paul Ehrlich, Jem Bendell, Donella Meadows, Paul Chefurka, Nate Hagens, Arthur Keller, Michael Dowd, Derrick Jensen, Jean-Marc Jancovici, and many others. Special thanks to Alexandra Elbakyan for her courage and initiative on open access to information.

Dear reader,

Thank you for buying this book. I hope you found value in it. Please consider rating or reviewing it on your favorite online bookstore. I'd like to know your honest opinion—it only takes a few seconds and may help others discover the book and decide if it's right for them.

Thank you—I read every review.
JP

Go to the website **bit.ly/m/collapse** or scan the QR code if you'd like to share a rating or review.

About the Author

Juan Pablo Quiñonez is a mestizo Latino who has been ruminating on the predicaments of modernity for over a decade. His writing aims to bridge ancestral and Indigenous perspectives, psychology, spirituality, resilience, systems thinking, science, and deep ecology, exploring where we are and our roles in facilitating what emerges.

He was born and raised in Guadalajara, Mexico. As a teenager, he completed basic training in the French Foreign Legion but left shortly after he received his parachutist badge. He studied at MRU in Canada, where he met his wife, Jennifer Ford. After graduating, they paddled into the boreal forest, where they spent six months foraging in isolation to supplement their minimal rations.

Juan Pablo has spent a hundred days in the forest foraging in the solitude of the boreal winter. In 2021, he won the ninth season of the reality TV series *Alone* after surviving solo for seventy-eight days in the subarctic lands of Labrador during the fall, where he underwent a multi-week water fast. In addition to being an amateur collapse researcher, he is also a wilderness survival expert specializing in the boreal forest and has written the survival book *Thrive*.

Drawing on research, life experiences, and insights, he urges us to challenge our assumptions and cultivate an openness to what is happening around us. Juan Pablo encourages us to stand on a deeper, more solid foundation, enhancing our ability to perceive uncomfortable thoughts and sensations without losing ourselves. He calls us to embrace the paradoxes, complexities, and magnitude of our predicaments.

He believes that the right insights will empower us to respond in our own ways with wisdom and courage, and they will guide us in letting go of what's unhelpful, so that a more beautiful world can emerge. Juan Pablo lives in the boreal forest in Canada.